AF477251

Aquaculture Technology

Aquaculture Technology

Hiralal Chaudhuri
A. B. Chaudhuri

2024

Daya Publishing House®

A Division of

Astral International Pvt. Ltd.

New Delhi – 110 002

First Published, 2009

Reprinted, 2024

ISBN: 9789359192062 (HB)

Disclaimer:

Published by   :   **Daya Publishing House®**
A Division of
Astral International Pvt. Ltd.
– ISO 9001:2015 Certified Company –
4736/23, Ansari Road, Darya Ganj
New Delhi-110 002
Ph. 011-43549197, 23278134
E-mail: info@astralint.com
Website: www.astralint.com

Acknowledgements

The editor thanks the Director, CIFRI, Barrackpore, Chief Editor, "Fishing Chimes", Editors "World Aquaculture", "Aquaculture Asia" and "Journal of Indian Fisheries Society" for allowing him access to their journals and utilize relevant contents under proper acknowledgements. Thanks are also due to Dr. S. B. Singh, former Advisor, FAO/UNDP for his encouragement and directions in preparation of the manuscript.

A.B. Chaudhuri

Dedication

To the millions of poor fisherman of India, Myanmar (Burma), Lao PDR, the Philippines and other countries with whom and for whom Professor Chaudhuri worked, sometimes in unhealthy habitats and insurgent infested areas; the professor has wished them his love, affection and regards.

To Late Mrs. Mukul Chaudhuri, wife of Prof. Chaudhuri, whose work at social level touched the hearts of millions poor fishermen communities; and for extending moral and financial support to the editor in the preparation of this manuscript.

A.B. Chaudhuri

Dr. P. Das
(Dr. Punyabrata Das)
Ph.D. FNASc, FZS, FBRS FNC, FSB, FAEB, FAA, FIFI
Specialist in Aquaculture Genetics (U.K.)
Retd. Director, National Bureau of Fish Genetic Resources, Lucknow, ICAR, Govt. of India,
Ex-Fishery Consultant, West Bengal University of Animal & Fishery Sciences, Kolkata
Member, W.B. State Agriculture Commission, Kolkata

Foreword

I feel thrilled in writing a few words on the book entitled *'Aquaculture Resurgence: Birth of Blue Revolution'* containing epoch making contributions of Prof. (Dr.) H. Chaudhuri, the Legendary personality in fishery science of the present world.

The success in Induced Breeding of Major Carps in confinement, standardisation and its application in varied eco-climatic conditions opened a new era in aquaculture of high production, as achieved now. The basic material itself for the modern aquaculture was the gift of the 'Father of Induced Breeding in India' who is a silent worker-cum-teacher of outstanding cordiality.

The present attempt of documenting the treatise of Prof. Chaudhuri's work in South and Southeast Asia in particular and the World in general would lead to further research by young Scientists in the multidirectional fishery Science.

I appreciate the effort of the Editor, Shri A.B. Chaudhuri, IFS, Ex-Director, Forest Survey of India for his present endeavour.

Dr. P. Das

A8/4, Indralok Estate, Paikpara,

Kolkata – 700 002

Preface

To the Readers

Since late 90's when some local news papers flashed the news of Professor Chaudhuri's success in induced breeding of carps and his monumental achievements in aquaculture resulting in profuse carp seed production, there had been a lot of queries about more detailed information about his successes episode. But neither the professor himself (he was averse to any publicity, name and fame) nor our Government gave the discoveries the importance it deserved.

Thereafter, felicitations and accolades came in profusion from various universities and institutions offering him their honorary membership and D.Sc (*Honoris causa*) and there was more public demand to know in details of the professor's contribution to aquaculture. Prof. Chaudhuri was highly honoured by three Universities of Japan and the USA for his outstanding work and World Aquaculture Society offered him lifetime honorary membership. At such a moment came a request from M/s Daya Publishing House that they were interested to publish Professor Hiralal Chaudhuri's lifetime contributions to aquaculture. It was thought a visionary and praiseworthy offer in the context of World wide recognition of his work.

The present editor was in a dilemma. It was because of Professor's apathy for any publicity of his work and his inability to help or untying various complicated knots in his papers for this write up if any. Another, apprehension was, if this editor (myself) was competent to present the Professor's work in true perspective without any misrepresentation of facts; and further that there could be wrong speculation that the professor was interested in the present publication although he was never behind this endeavour. Prof. Chaudhuri did not need any more recognition to his successes which he has already achieved.

The professor very reluctantly acceded to the request of this editor which the readers may peruse in the letter appended hereinafter.

On being successful in getting his consent the editor collected reprints, reports and journals those were available at his Saltlake residence. This treatise like two others now under publication by the same Publisher does not present all the achievements of the professor as the bulk of his work donated to the libraries of ICLARM and University of the Philippines was not available for analysis and presentation.

As such, this endeavour is primarily meant to bring to highlight Prof. Chaudhuri's contribution to aquaculture in South and Southeast Asia and designed to satisfy the need of those mostly non-technical and non-aquaculture enthusiasts. The Professor's magnanimity to consider his successes as joint effort of those who were associated with him was well-known.

The editor shoulders all shortcomings for his inability to present Prof. Chaudhuri's work in proper flavour without any blemish as his work was highly technical and complicated. The editor also begs of the readers his unqualified apology for any mis-statement, mistakes or mis-representation of facts presented by him.

In the initial stages of his research endeavour the Professor was helped by late G. N. Mitra, Director of Fisheries, Govt. of Orissa by extending research facilities; at research station he was helped by Dr. K. H. Alikunhi, Head of research station, Cuttack and he was grateful to both of them. The professor acknowledges with all sincerity the help and assistance extended by Dr. Srinivasa Rao, Dr. C. V. Kulkarni, and Dr. B. R. Bhatia and he thinks all his successes were possible because of their active help.

For proper evaluation of Prof. Chaudhuri's work in India, South and Southeast Asia especially by the non-technical and non-aquaculturist readers, the editor has presented the following observations in Chapter 1:

1. Stalwart aquaculturists assessment of his work
2. Forecast of a brilliant era of aquaculture in Southeast Asia
3. Professor's contribution to South and Southeast Asia in particular as viewed by some Universities and World organizations.

These views should be considered enough to evaluate the Professor's contribution to aquaculture in general and especially for the people of India, South and Southeast Asia.

An Amazing Discovery

The fifth decade of 20th Century witnessed aquacultural resurgence in the discovery of "Induced Breeding" of major carps. July 10, 1957 was a red-letter day in aquaculture research. Revelation of life history of milkfish (*Chanos chanos*) in 1977 which made induced breeding of this fish a success was another land-mark of his discovery. Earlier, he was given the epithet as "Father of Induced Breeding" for his series of successes in induced breeding.

Well-deserved rare honour given to Prof. Chaudhuri was the crowning glory of his outstanding works. His findings made possible the planned fresh water aquaculture of today in respect of induced breeding of carps by the administration of pituitary hormones, the most important scientific methodology developed in the field of cultured fisheries. This success made it possible to obtain good quality seeds. He developed high yielding carp culture techniques in late sixties. He also demonstrated the practicability of achieving high fish yield in ponds by carps polyculture (Composite fish culture). His pioneering efforts in the use of reproductive hormones in finfish aquaculture and for his training of generation of students was highly recognized. The president of the Philippines late Fardinand Marcos congratulated him personally for his successful work of milkfish in 1977.

All these works gave an unprecedented boost to aquaculture research and brought a protein (blue) revolution in the World on a sustained basis.

The readers, it is hoped, will find full justification with the title of this treatise primarily from the observation of stalwart aquabiologists, some journals and institutions as recorded in nutshell in Chapters 1 and 2; Chapter 3 gives in short the professor's lifelong activities. Chapter 4 gives some views of aquaculturists on the scenario of 21st Century. Other chapters deal with some project works besides a few publications of the professor.

So having spent second half of his active career in Burma, Lao PDR and the Philippines and rest (first half) in India, the professor's lifetime success and the fruits thereof pertain to the people of India, South and Southeast Asia.

A.B. Chaudhuri, IFS (Retd)
131, NSC Bose Road
Block – 10, Flat – 4,
Kolkata – 700 040
Ph: 91-033–2471 6734

Professor Chaudhuri's Consent for Publication

Although I have always been averse to name, fame and public appreciation for whatever little I have achieved in my Scientific endeavour, I rather reluctantly permit Shri A.B. Chaudhuri of 131, N.S.C. Bose Road, Block-10, Flat-4, Kolkata – 700 040 to edit my research documents, lecture notes, reports etc. relating to aquaculture, limnology and ichthyology either in part or full or modify them for suitable publications.

All these papers were widely read discussed, quoted and presented in national and international Scientific Workshops, Symposia and Congregations. As such no accent is necessary for present publication from any body or any organization.

Kolkata

Sd/-
H. Chaudhuri
Formerly Chief Technical Advisor, FAO/UNDP and
Regional Aquaculture Coordinator
South East Asian Fisheries Development Centre, Manila

Contents

Chapter 1

Aquaculture Resurgence: Birth of Blue Revolution

A RESUME

This chapter unsuitably records the stamp of success of Prof. Chaudhuri in India, South and Southeast Asia. A galaxy of bio-aquaculturists and top International institutions have acknowledged his outstanding success in various facets of aquaculture.

"Aquaculture Asia", a journal of eminence accepts the professors "Induced Breeding and mass seed Production" as the major contribution to Southeast Asian Aquaculture

World Aquaculture Society (WAS) observes:

Under the auspices of the FAO, this expertise in freshwater fish breeding and grow-out has been shared with numerous extension workers and other government staff in many countries, including his own (India), Laos, Myanmar, the former Soviet Union, Malaysia and Fiji. He also managed a large FAO field project in Laos for five years from 1979 to 1984. In his post in the Philippines Hiralal Chaudhuri worked in a collaborative training program developed between SEAFDEC itself and NACA, the Network of Aquaculture Centres in Asia, which has its base in Bangkok, Thailand.

WAS say he is truly an international figure and is rightly called the "Father of Induced Breeding". Outstanding Contribution of this scientist made possible planned freshwater aquaculture in Induced breeding. His pioneering research since 1950's made possible to obtain seeds of good quality on a sustained basis.

This chapter, is in fact, justifies the title of the book.

Every fibre of Professor Hiralal Chaudhuri's mortal being was tuned to the development of aquaculture not only in the South and Southeast Asia, but also all over the World. Resurgence, in fact started on 10th July 1957 with the success in his induced breeding of Carps in Cuttack. The success continued upto 1975 as a researcher with Govt. of India and, thereafter from 1976 to 1993 with SEAFDEC, FAO/UNDP and the Philippines University in various capacities. The successes were achieved wherever he laid his hands. Blue (Protein) Revolution was a corollary to such successes. Some facts are presented hereinafter to substantiate the aforesaid observations.

A. Unique and Outstanding Achievement: "Aquaculture Asia" Views

The Journal "Aquaculture Asia" asked two expert aquaculturists and veteran Aquaculture Development experts what they thought had been one of the major factors to Asian Aquaculture Department.

The response was:

"Induced breeding of carps and their mass seed productions".

[Source: *Mr. Nisheet Bhatt, Aquaculture Asia's*
Technical Correspondence reports in the Journal
Vol.V No. January–February 2000]

None other than Prof. H. Chaudhuri deserves this appreciation. The first success obtained in 1957 continued long years in various fields in South and Southeast Asia and other parts of the World.

B. Technological Break through of the Millennium

Dr. P.V. Dehadrai in his lecture on Prof. Hiralal Chaudhuri Fisheries Foundation held in Versova, Mumbai in 2001 says;

"10th July 1957 was an important day for Indian Aquaculture. It was Technology break through of the Millennium when Rohu was induced breed by hypophysation. The hero was Dr. Hiralal Chaudhuri in Collaboration of Shri K. H. Alikun at CFRI Research Station at Cuttack's Killa fish farm. It set the trend for breeding several commercially important food fishes as well as ornamental fishes.

However, the World over efforts were already going on to understand the mechanism of hypophysation. It took almost 2 or 3 decades to understand the semantics of reproductive hormonal interactions in relations to the interval rhythm of the ambient environment both aquatic and atmosphere".

Professor Chaudhuri studied thoroughly the seasonal reproductive cycle maintained by endocrine cycle and how the hormonal cascade has perfectly coordinated with seasonal reproduction cycle of the fish to ensure spawning at a specific time of the year. He understood how the environmental factors regulate reproduction by the endocrine or hormonal path way.

C. World Conferences and Symposia–Views

The symposium on "Warm water fish pond culture" held in Rome in May 1966 recorded various activities relating to the subject. Since then some regional conferences were held in various countries. But the landmark was FAO sponsored "World Aquaculture Conference" held in Kyoto, Japan in 1976. It recorded a comprehensive and consolidated step to record the progress of various facets of Aquaculture. It concluded that "aquaculture has developed to become the fastest growing food producing sector in the World; it 'has expanded, diversified, intensified and technologically advanced and its contribution to aquatic food production has increased significantly. Aquaculture is highly diverse with a broad spectrum of systems,' policies and operations ranging from backyard small, household pond systems to large scale highly intensive practices (modified from Aquaculture Asia, 2000).

Bangkok Declaration and strategy for aquaculture development beyond 2000, made at a conference in 2000 provide a blue print for Government action for aquaculture development. The Journal "Aquaculture Asia" (Volume, Oct-Dec. 2000) records that Asian aquaculture has progressed a lot since 1980. Kyoto conference while making the declaration records the development of aquaculture in the past decade.

Conferences and symposia have not specificably mentioned Professor Chaudhuri's epoc-making discoveries and over-looked thumping appreciations offered to him by many world aquabiologists from many research institutes made on his success in induced breeding and in other fields of aquaculture. This editor has however presented scores of facts to establish and confirm that aquaculture got a boost in the success of induced breeding in 1957 and quickly went on gaining momentum.

The professor's foreign assignments in Burma, Lao PDR, SEAFDAC, and in many other countries proved immensely successful bringing revolutionary improvements in various facets of aquaculture in those countries. As such, aquaculture no doubt rejuvenated since 1957 in South and Southeast Asia and the World.

Elliot Enlis (World Aquaculture, March 1997) uses a Term "Aqua biotech"–a blue revolution to record a profound increase in productivity widely heralded as the green revolution. India had such green revolution in 50's, 60's and 70's that doubled crop yield in a short line. With the application of discoveries from biotechnology South and Southeast Asia region witnessed a similar leap in the fish protein productivity since 1990.

This statement is nullified by the facts and figures presented in this chapter (also in other chapters) by this editor that blue revolution started in 1957 with the discovery of induced breeding. This discovery gained momentum in various other discoveries in subsequent years culminating in pinnacle of successes by the end of last millennium.

D. Resurgence and Blue Revolution takes off on July 10, 1957 in Successful Induced Breeding

The editor records a few thumping felicitations offered to Prof. Chaudhuri from top world famous aquaculturists on his success in induced breeding. This chronological successes in induced breeding also of several other species, profuse seed production, improved nursery practice, composite fish culture, cross breeding, training of personnel make an epoc-making event that has benefited many countries in South and Southeast Asia.

Some experts have designated this success equivalent to getting the Nobel Prize; some say the work is "as glorious as a golden page in the annals of Indian Fisheries'; others say he has created a fish-cultural history.

The work of the Prof. Chaudhuri gained momentum on his appointment as FAO/UNDP expert on Fishery Development in Burma in 1966 and again from 1967–70, and month long tour in European Countries on "Genetic selection and hybridization of cultivated fish (1968)" and assignments in Malaysia in 1974 and

1975; later after joining SEAFDEC in 1976 he worked in Malaysia again (1978), Sudan (1978), Lao PDR (1974-79), Fiji (1974), Nigeria (1978) and elsewhere for improvement of various facets of research and training in Aquaculture.

Chapter 2

Contribution of Professor Hiralal Chaudhuri to Aquaculture: Views on Resurgence and Blue Revolution in South and Southeast Asia

A RESUME

To avoid any divergence of opinion or controversy about the professor's contribution to aquaculture resurgence and blue revolution the editor presents views of a good number of world famous aquaculturists and of some institutions which emphatically confirm that Hiralal's contribution was meteoric and various facets of aquaculture got tremendous boost in all his research successes.

The presentation has been made in five parts viz.

A. On Induce breeding

B. On his posting in SEAFDEC

C. Confidence on success of milk fish breeding.

D. Other Institutions

E. Fish Farmers' day

A. Induce Breeding: World Expert Aquaculturists Views

Success in Induced Breeding of Carps initiated a resurgence and blue revolution in aquaculture in 1957.

☆ "Congratulation on your success in induced Spawning".

(Dr. Stillman Wright, USA Department of Interior, 1957).

☆ "The best student ever to enroll in Fishery in the Institute"

(Dr. Swingle, Alabama Institute, USA, 1957)

An authority on Aquaculture.

☆ "Chaudhuri, if you succeed in induced breeding of Carps, it will be comparable similar to getting the Nobel Prize".

(Dr. B. R. Bhatia, Deputy Fisheries Adviser, Govt. of India, 1957)

☆ "You are making Fish Cultural History. You are making an important contribution to economies".

(Dr. J.W. Atz, USA, 1958)

☆ *Dear Dr. Chaudhuri,*

Congratulation to you on your successful experiments in inducing spawning in carps by means of hormone injections. I trust that you will be able to transfer the results of your laboratory work to the practical application on carps throughout the country, and thus solve the problem presented by failure to spawn naturally.

I take this opportunity to wish you a Happy New Year in 1958, and continued success in your work. You will be interested to know that I had a recent conversation with Dr. Swingle and that he was extremely complementary of your work at the Alabama Polytechnic Institute. He told me that you were one of the best students ever to enroll in fishery work at the Institute. This statement is gratifying to me as I am sure it would be to you.

Dr. Stillman Wright, Chief,
United States Department of Interior Office of Foreign
Activities Fish and Wildlife Service, Washington–25, DC
December 27, 1957

☆ *Dear Dr. Chaudhuri,*

First of all, I want to congratulate you on your success in inducing freshwater food-fishes of India to spawn in captivity. You are certainly making fish cultural history, and in view of the tremendous importance of fish have in human providing animal protein food, the item in man's diet most critically in demand; you are also making an important contribution of economics.

I should be very happy if Dr. Pickford and I could speak of your success in our forthcoming article in ENDEAVOUR. As you undoubtedly know, this journal has a very wide distribution and I am sure it would make your fine work known to many people who might otherwise remain ignorant of it............

Sincerely yours,
Dr. James W. Atz
New York Zoological Society,
July 23, 1958

☆ *Dear Dr. Chaudhuri,*

Promising developments have recently occurred in India. The principal problem in Indian Fish culture has been obtaining seed fish regularly. H. Chaudhuri of Central Fisheries Cuttack has now been able to induce spawning in six species of carps..."(Dr Pickford and Dr. Atz are world famous fish Physiologists. They are the authors of the famous book "Physiology of Pituitary Gland of Fishes").

From: "ENDEAVOUR"
By J.W. Atz and G.E. Pickford
"The use of pituitary hormones in fish culture".
July, 1959

☆ *Dear Dr. Chaudhuri,*

..............I want to thank you for all your kindness and help to me during my visit. And that goes for your colleagues as well I think you are doing wonders with the inadequate equipment you have to work with, and you set an example for us."

(Dr. C.F. Hickling, Director, Trop. Fish Culture Institute, Malacca and Eminent Fish Culturist Research Institute 1965)

☆ *Dear Dr. Chaudhuri,*

I wanted to tell you how much we enjoyed our visit to your fisheries research station. It was an eye opener to us all, amply confirmed when we repeated our visit in West Bengal.

As Chairman of the Technical Advisory Council I was very interested because of the attention we have been giving to aquaculture and also as a Governor of the IDRC I was doubly interested. You furthered my education very much and I offer my warm congratulations on your success.

With good wishes,
Yours sincerely,
Sd/- Dr. J. G. Grawford
Governor, World Bank
1974

☆ *Dear Dr. Chaudhuri,*

I was shocked to learn from your letter that you had opted voluntary retirement. Your association with important researches in Inland Fisheries and especially your contribution in respect of induced breeding of carps and also intensive fish culture has undoubtedly been quite monumental. These works will remain as glorious as a Golden Page in the annals in the Indian Fisheries research in India. You were very rightly awarded the covetous Rafi Ahmed Kidwai Memorial Prize. I once again congratulate on that meritorious achievement. Nevertheless, you have retired at a time when our country needed you most. I do not know whether there are chances of rethinking. However, I trust that you will continue the work in your beloved field of Inland Fisheries. I earnestly wish you further laurels and happy long life.

Dr. D.V. Kulkarni
Indian Fisheries Scientist & Administrator
Director of Fisheres of Maharashtra (Retired)
Bombay, India
March 16, 1976

[Dr. Kulkarni is a renowned Fisheries Scientist and an administrator of India. He retired as the Director of Fisheries, Bombay and held many high positions. This letter was written after hearing from Dr. Chaudhuri about his voluntary retirement from ICAR· Service]

☆ It is on record that in 1973 Prof. Chaudhuri was offered a very high post by FAO/UNDP to work in Colombia (Latin America) for advice on Aquaculture development in several countries in Latin America. But he could not get release from the Govt. of India.

B. Confidence on Prof. Chaudhuri about Aquaculture Revolution on his appointment in SEAFDEC

Prof. Chaudhuri resigned from India Govt. service in February 1976. Immediately after, he got an offer from SEAFDEC to join as Regional Aquaculture Coordinator to monitor the Aquaculture programmes as Deputy Director of Aquaculture.

Later, during 1979–1984, he joined as the project coordinator / Project Manager/Chief Technical Advisor of Aquaculture Development project at Vientiane. On expiry of the period he once again joined SEAFDEC in 1985 as Senior Technical Advisor for 4 years.

The editor presents happy reactions of a few (out of many famous aquaculturists on his appointment in SEAFDEC which prove how emphatically they lauded his appointment and were confident about Aquaculture in South and Southeast Asia would get a tremendous boost on his new appointment.

These records themselves prove that Prof. Chaudhuri's indispensable series of success in South and Southeast Asia.

☆ Dr. K. Kuronuma, President of Tokyo University designated Prof. Chaudhuri as early as in 1975 as the "Father of Induced Breeding"

☆ "Your appointment in SEAFDEC is a cause of considerable pleasure, to others in this country as well as myself".

(Dr. G. R. Fish, Fishery Research Lab.,)

☆ *Dear Dr. Chaudhuri,*

I was very pleased to hear from you and I would like to extend my congratulations to you on your new appointment in SEAFDEC in the Philippines.

Your appointment is cause of considerable pleasure to others in this country as well as myself..............

I have sent a copy of the full report that our Freshwater Sciences Committee made to the Pacific Science Association to you under separate cover. I am most grateful to you and to the other members of the Committee for the help I received during our term of office.

Sd/- G. R. Fish
Fisheries Research Laboratory
P.O. Nagapuna, Rotoura
New Zealand
6 May, 1976

☆ "Your services can be provided for coordinating the stimulated development of aquaculture"...."Your duties with SEAFDEC" may help catalize on harmonious development activities in Southeast Asia".

(Dr. W. H. L. Allsop, Canadian Director, Aquaculture, 1976)

☆ *Dear Dr. Chaudhuri,*

I wish to congratulate you on your new position with the Southeast Asia Fisheries Development Centre. I am sure you will find this a challenging experience, and I wish you every success.

Sd/- Lauren R. Donaldson
World Famous Fishery Scientist and Professor Emeritus of Fisheries
University of Washington
College of Fisheries
204 Fisheries Centre
Seattle, Washington 98195
May 20, 1976

☆ *Dear Dr. Chaudhuri,*

I was delighted to hear of your receipt of the Rafi Ahmed Kidwai Memorial Prize, which you thoroughly deserve. I continue to sing your praises of the very fine work done by you and your colleagues in developing aquaculture.

It is no surprise to me that you should have been invited to join the Southeast Asian Fisheries Development Centre and I warmly wish you well in that new responsibility–much as you will be missed in India.

Sir John Crawford,
Professor and Chairman
Technical Advisory Council Aquaculture (1976)

☆ *Dear Dr. Chaudhuri,*

The intention of SEAFDEC together with the Southeast Asian Centre for Graduate Training and Research (SEARCE) is that through their joint auspices, and the partial funding from the Milkfish Project (IDRC), your services can be provided for coordinating the stimulated development of aquaculture programms in the counties affiliated to SEAFDEC and SEARCA.

May I indicate a few comments in regard to how we anticipate your duties with SEAFDEC may help catalyze on aquaculture programme. It is anticipated that your long and harmonious association with aquaculture research and development activities in Southeast Asia (including, I believe, Burma, Fiji, Malaysia, Indonesia, Bangladesh besides your home country of India), will enable you to assist in the continued promotion of such programs with the assistance of SEAFDEC and SEARCA....................

You have already earned the respects and compliments of many projects in different countries, now you have a bigger fish pond to swim in.

Yours sincerely,
Sd/- WHL Allsopp
Canadian Aquaculturist,
Head: IDRC Fisheries
Associate Director (Fisheries)
International Development Research Centre,
Otawa, Canada
March 29, 1976

[The above letter was written just after Dr. Chaudhuri's voluntary retirement, when the offer came to join the position as the Regional Aquaculture Coordinator. Although he was appointed by SEAFDEC, Dr. Chaudhuri was intended to monitor also the projects to SEARCA and IDRC Aquaculture Projects.]

C. SEAFDEC Posting: Prediction About Success of Induced Breeding *Chanos chanos* (Milkfish)

Prof. Chaudhuri joined SEAFDEC in early 1976 and found that frantic efforts were made by various experts working in Hawaii, Tiwan, Philippines and Indonesia to make induced breeding of *Chanos chanos* a success. However, Prof. Chaudhuri was not concerned with direct research on the fish as IDRC, had already engaged a team of scientists to work on the fish, and he was only the coordinator.

The editor presents the observations of a few top Aquaculturists who predicted as early as in 1957 and thereafter that such a difficult task of induced breeding of *Chanos chanos*, could perhaps be done successfully by Prof. Chaudhuri (which is a superlative confidence and honour to him) which are as follows.

☆ "If anybody could accomplish this difficult task (of induce breeding of milk fish) Dr. Chaudhuri would be one".

(Dr. D. D. Moss, (USA Aquaculturist)

☆ *Dear Dr. Chaudhuri,*

Your letter of Dec. 16, 1957 was very exciting to me. For several years since my visit to the Far East I am waiting for the day when Scientists from that part of the world will succeed to induced spawning of some of the important fish of that region.

It seems to me that if you succeeded with the Indian carps then, why not follow with other species like Chinese carps, Mullets and *Chanos chanos?* This might be a big revolution in biology and fish culture as well. So, I wish you much success and progress. Meanwhile I am waiting for your publications on these Experiments.

Sincerely yours,
Sd/- S. Sarig
Editor- "BAMIDGEH"
(Israeli Journal of Fisheries and Aquaculture)
Nir-David, D. N. ISRAEL
December 27, 1957

(NB: Dr. Shmuel Sarig is one of the few TOP Fishery Scientists of Israel and has earned World-wide recognition as an Aquaculturist and Fish Pathologist as well as an Editor)

☆ "If you succeed in Induced Breeding of Milk fish it might be a big revolution in biology.

(Prof. Chaudhuri work on *Chanos chanos* has been presented separately.)

(Dr. S. Sarig, Isreal, 1958)

☆ *Dear Dr. Chaudhuri,*

We have been keeping up with your progress in breeding the milkfish. With some visitors the other day, Dr. Shell* made a comment that if anybody could accomplish this difficult task that one Dr. Chaudhuri would be the one. Please do keep us informed of your milkfish particularly when you have been successful in not only producing the larvae.............. In reference to publishing your work on the milkfish I would suggest that you contact two journals:

   1. Aquaculture   2. Fish Farming International

Sd/- D. D. Moss
Famous U. S. Aquaculturist
Assistant Director,
International Centre for Aquaculture
Department of Fisheries and Allied Aquaculture
Auburn University
Auburn, AL 36849-5419
June 30, 1977.

*[*NB: Dr. E.W. Shell is the Prof. and Head, Department of Fisheries and Allied Aquacultures and Director, International Centre for Aquaculture.*

The paper "Observations on artificial fertilization of eggs and the embryonic and larval development of "milkfish Chanos chanos (Forskal)" by H. Chaudhuri et al. was published in "AQUACULTURE", 13 (1978), 95-113 (Printed in Netherlands)]

☆ *Dear Dr. Chaudhuri,*

It is very nice that the eggs of *Chanos* which were collected undoubtedly under confinement have been grown to post Larval stage this fact itself, the world record in strict sense, is a significant step towards the goal. I am very glad to find this record established in the Philippines before Hawaii and Taiwan.

..................As to the publication of your valuable contribution in early date, I strongly suggest it done by SEAFDEC......................... SEAFDEC is an International Organization whose contribution should not be kept under national level.

Sexual dimorphism of *Chanos* through your hand and Biology was made clear.

Sd/- Katsuzo Kuronuma
Eminent Ichthyologist and Fisheries Scientist of Japan
June 29, 1977.

[NB: The letter was written after Dr. Chaudhuri's success in artificially fertilizing milkfish (Chanos) eggs caught from the sea and studying the development and rearing upto the post larval stage. He published the paper in the Journal 'Aquaculture' from Netherlands. He had also succeeded in determining the sexes of milkfish (Identifying characters).]

D. Other World Institutions Felicitation

The contributions of Prof. Chaudhuri in the development of aquaculture in India, Southern Southeast Asia has been very precisely expressed in a few felicitations which several universities and World Aquaculture Organizations offered him. These rare honour extended to him for life long service to Aquaculture and the people testify to his work in resurgence and blue revolution in the arena of aquaculture.

Blue (Protein) Revolution: Views of World Organizations

The fifth decade of last Century witnessed a "Green Revolution" when the agricultural research augmented the production of food yield considerably. The next decade saw the foresters' initiating a "Wood Revolution" by introducing Eucalyptus, Akasmoni and other fast growing species on sandy, rocky and almost sterile soil and bringing vast eroded areas under sylvan cover. Professor Chaudhuri's research success in 1957 and onwards brought about a "Blue Revolution" or a "Protein Revolution" which very rapidly created a vast fish protein resource base on a sustained basis; he was able to reach million tonnes of fish larvae to the village fishermen for fish farming. Millions of poor people of this and other countries now have an easy source of protein diet following his research findings.

In words of Prof. H. Chaudhuri "Aquaculture development has reached to such a shape that it is now capable of making a significant contribution to the supply of fish protein to million's of poor people of the world suffering from malnutrition due to deficiency of animal protein in their diets. With further intensive research on the biological, endocrinological, nutritional, environmental and socio-economic aspects and infrastructures confined with massive financial investments and suitable legislation, it will be possible to obtain unbelievable production of this animal protein for the benefit of mankind"

Prof. Chaudhuri thought that aquaculture research and its success has attained such a high level that fish protein would meet the needs of the poor people and thus save them from malnutritional diseases.

Tributes by three Universities and World Aquaculture Organization

This tribute itself is a glorious example to the professor's contribution to aquaculture in India, Burma, Lao PDR, the Philippines and other Countries.

In September 8-11, 1994, a colloquium on "Application of Endocrinology to Pacific Rim Aquaculture" was organized by the Fish Endocrinologists/Aquaculturists of three prestigious Universities, namely University of California at Berkeley, University of California at Davis and University of Tokyo, Japan, in Honour of Prof. Hiralal Chaudhuri at Bodega Marine Lab., USA. The proceedings of the Colloquium published in a special issue of AQUACULTURE journal, was dedicated to him in recognition of his pioneering effort in the use of reproductive hormones in finfish aquaculture and for training of generations of students in India, Myanmar, Laos and the Philippines. He was commended his sustained international contributions over five decades and for his talent in applying the results of basic research to the solution of practical problems.

A Special Issue of the Journal "World Aquaculture" (1995) Reads:

"This special issue is comprised of the papers and abstracts presented a colloquium held in honour of Prof. Hiralal Chaudhuri at the University of California Bodega Marine Laboratory on 07-11

September 1994. This was the seventh in a series of annual conferences held at the Laboratory.

The colloquium was organized by:

Sd/- A. Bern, Howard University of California, Berkeley,

Sd/- Ernest S. Chang, University of California, Davis,

Sd/- Tetsuya Hirano, University of Tokyo.

This topic was chosen because all Pacific Rim countries have an increasing interest in the practice of aquaculture to supplement food production. Owing to intense fishing pressures, the aquatic culture of shellfish and finfish plays and will continue to play a major role in the economics and research thrusts in most of these Pacific Rim countries. One area of aquaculture research that is currently experiencing intense effort is hormone manipulation. Because we had to keep the meeting small and focused we were constrained to limit North American participation in the meeting to colleagues in the western provinces and states of Canada and the USA.

The papers were divided into section on reproduction, development, salmon smoltification, growth, metabolism and general endocrinology. There was also an evening workshop on maturation and reproduction in captive salmonid populations. The meeting stressed a comparative approach and included papers on invertebrate and vertebrate species of interest to aquaculture.

Professor Hiralal Chaudhuri, the participating guest of honour at this meeting, receives our grateful recognition for his pioneer efforts in the use of reproductive hormones in finfish aquaculture and for his training of generations of students in India, Laos, Myanmar (Burma) and the Philippines. Professor Chaudhuri received his doctoral degree in Zoology from the University of Calcutta in 1961 and has provided counsel to the aquaculture programs of many countries, including India, the Philippines, Laos, Myanmar, the formmer Soviet Union, Indonesia, Malaysia, Thailand and Fiji under the auspices of the Food and Agriculture Organization of the United Nations. He is truly an international figure. Professor Chaudhuri has been called the "Father of Induced Breeding" his pioneer study on the use of a pituitary preparation to induce spawning in carps was published in 1957. He has recently published a history of induced breeding in fish in the Proceedings of the Zoological Society of Calcutta [47 (1): 1-31]. He retired from the University of the Philippines to his home in Calcutta in 1993, after 11 years of research and teaching in the Philippines."

Resurgence of Aquaculture took a start in 1957 Carp Breeding and Fish Seed Production: An Acknowledgement

This well-deserved rare honour to Dr. Chaudhuri is the crowning glory of an outstanding scientist whose contributions have made possible the planned fresh water aquaculture of today in respect of induced breeding of carps by the administration of fish Pituitary hormones, the most important scientific methodology development in the field of culture fisheries. This pioneering research work done by him in the fifties made possible to obtain seed of good quality instead of the earlier method of collection mixed seed (desirable and undesirable fish species together) from natural sources. His work made possible the raising of unmixed fish seed in farm ponds and subsequent stocking of waters with the kind of fish seed desired.

Dr. Chaudhuri did not rest with the development of this scientific method of breeding fish but carried on to take this to its logical conclusion i.e., optimising production from fish ponds. He developed high yielding carp culture techniques in the late sixties, the first outstanding results obtained in Burma when serving as a FAO/ UNDP Inland Fishery Biologist, he achieved high production (10 tonnes/ha) of carps in fish ponds. Later, in return to India he demonstrated the practicability of achieving high fish yields in ponds by carp polyculture (Composite fish culture) through applied research work in Cuttack at the Central Inland Fisheries Research Institute, the same Institute where he had developed the induced fish breeding technique.

The excellent instructional film "Induced Fish Breeding" produced by the Films Division of the Government of India was made at Cuttack under Dr. Chaudhuri's technical guidance. The film had received national honour as also the award for the best film at the FAO sponsored symposium on Aquaculture in Kyoto Japan in 1976.

Dr. Chaudhuri combines in him the rare abilities of an excellent teacher the profession he started with after his brilliant academic career at the University of Calcutta as the outstanding researcher and the highly successful extension scientist. After joining the newly formed Central Inland Fisheries Research Institute in the late forties, Dr. Chaudhuri was sent in the mid-fifties on deputation to the USA, to work under Fish and Wildlife Service, by the Government of India. There he obtained the M.S.

In recognition of his outstanding contribution in the field of fisheries and aquaculture he was offered many consultancies/ assignments by the FAO, IDRC, ESCAO, WORLD BANK, SEARCA and SEAFDEC to work for the development of fisheries and aquaculture and training personnel in a number of developing countries of the world.

He served the FAO/UNDP as a fisheries expert for about 10 years working in Burma and Lao PDR. He was the Regional aquaculture co-ordinator of the South East Asian Fisheries Development Centre (SEAFDEC), Philippines, as also its Senior Technical Adviser and visiting Scientist for about 7 years and made significant contributions towards the success in breeding of the milkfish, *Chanos* besides training a host of visiting fisheries officials form different countries.

Dr. Chaudhuri was visiting professor at the University of the Philippines at Los Banos till recently, from where he retired after about 5 years.

A unique appreciation came his way in 1994 when the University of Calcutta, his *alma mater*, instituted a chair in Fisheries in Chaudhuri's name.

Inland aquaculture owes a permanent debt to Dr. Hiralal Chaudhuri, for making possible through his painstaking brilliant efforts production of quality fish seed and how best to culture them through scientific farming to obtain high yields of food fish.

Life Membership of World Aquaculture Society (WAS) (February 1998)

The award goes at Dr. Hiralal Chaudhuri, who was born in Assam, India in 1921. He received his bachelor's master's and, finally in 1961 his doctorate in zoology from the University of Calcutta. In 1955 he also received an MS in Fisheries Management from Auburn University from his thesis on the effect of pituitary injections on pond fish reproduction.

For nearly 30 years until 1976, he was a researcher at the Central Inland Fisheries Research Institute in Cuttack, India where he made his greatest contributions to aquaculture through the hormonal induction of breeding in the Indian major carps and exotic Chinese carps, hybridization of cultured carps, improved nursery management techniques and developed the polyculture of Indian and Chinese carps.

He held a senior post in the Aquaculture Department of SEAFDEC (the Southeast Asia Fisheries Development Centre) in Iloilo, Philippines from 1976-1979 and the returned there in 1985-1988 as a Visiting Scientist. Under the auspices of the FAO, his expertise in freshwater fish breeding and grow-out has been shared with numerous extension workers and other government staff in many countries, including his own (India), Laos, Myanmar, the former Soviet Union, Malaysia and Fiji. He also managed a large FAO field project in Laos for five years from 1979 to 1984. In his post in the Philippines Hiralal Chaudhuri worked in a collaborative training program developed between SEAFDEC itself and NACA, the Network of Aquaculture Centres in Asia, which has its base in Bangkok, Thailand.

After visiting professorship at the University of Philippines for five years, Hiralal retired in 1993 to enjoy family life in Calcutta. About a year ago, he was awarded an Honorary Doctorate at the Burdwan University in West Bengal, in recognition of a lifetime devoted to aquaculture research and training. It is my pleasure to add to the recognition which his work has received by awarding Dr. Hiralal Chaudhuri honorary Life membership of the World Aquaculture Society.

E. Significance of 10th July, 1957–Fish Farmer's Day Spread of Aquaculture Consciousness

Dr. Hiralal Chaudhuri gifted to the nation and to the others on the globe a field oriented induced fish breeding technology developed by him on 10th July, 1957. This technology revolutionized fish culture operations leading to increase in fish production from fish farms on the one hand and virtual stoppage of spawn collections from rivers.

In recognition of this revolutionary contribution that provided a key to the up gradation of the socio-economic status of fish farmers, the Government of India declared 10th July, as Fish Farmer's Day for observing every year, from the year 2001. Since then all the states of India are celebrating this date as a 'Red–letter" day in the arena of aquaculture when the scientists assemble and discuss about the progress of aquaculture in various fields.

Dr. J.V.S Dixitulu, Chief Editor, Fishery Chimes Observes

"This event was the glorious success achieved by Prof. (Dr.) Hiralal Chaudhuri, Fishery Scientist of CICFRI at that time, in inducing major carps to breed through administration of homogenized harmone extract of fish pituitary into their body intravenously. This technology spread among fish farmers swiftly, thereby enabling the substitution of river carp spawn / fry collection practiced by them with this technology, which enabled them to produce seed of major carp species they wanted to culture, in numbers stocking in ponds from wild collections. The quantity of culture fish production cascaded over years to an enviable peak which has elevated India to an exalted status of being No. 2 in global culture fish production, next only to China. The upsurge in culture production has enhanced the per capita consumption of fish from an insignificant level in 1957 to an estimated 9 kg. per annum as at present. This remarkable development is the result of the devoted contribution made by a large number of aquaculturists.

Hiralal Chaudhuri is internationally acclaimed as the "Father of Induced Breeding" in recognition of his pioneering work on the breeding of Indian major carps and other fishes using pituitary hormones. He also successfully developed a number of technologies such as seed production, selective breeding, hybridization, hatchery management and integrated fish farming. He is virtually a living legend in aquaculture in India".

Fishing Chimes (January/February, 2000) Further Observes

"One distinct event of the century in the fisheries sector was the introduction and spread of induced fish breeding system adopting hypophysation technology. It imparted a conspicuous economic uplift to the culture fisheries sector of India. In this context, it will be pertinent to mention that the nation is indebted in fully measure to Hiralal Chaudhuri and K. H. Alikunhi and of course to several other scientists for the revolution that took place in the culture fishery sector because of this technology. The record of events related to the aforesaid revolution will be incomplete without a mention of HCG which came into use in the heels of the practice of administering pituitary hormone extract, followed by the later introduction of Ovaprim, a synthetic hormone product by Glaxo Company which obtained the product from M/s Syndel laboratories, Canada. This introduction was followed in recent years by the release of an indigenously manufactured similar synthetic hormone product known in the market as Ovatide. Ovatide is manufactured and sold by Hemmo Pharma and this product too has come to be accepted and used widely by fish breeders".

Chapter 3

Chronological History of Professor Chaudhuri's Activities

A RESUME

To help the readers assess the quantum of success in aquaculture research of Prof. Chaudhuri the editor presents briefly his life–time activities in chronological order.

This chapter clearly states that of 46 years of active services, Prof. Chaudhuri spent 28 years in Southeast Asia countries for his research contribution.

1957

Induced breeding among Asiatic Carp was a hard test of science. He achieved pioneering success in the world for application of pituitary hormone in *Chirrhinus reba* on 10th July in 1957. He also attained success among *Puntius sarana*, *Chirrhinas mrigala and Labeo rohita* species.

1957-59

A large number of trainees were trained up in Cuttack in these two years. Even foreign scientists came for training in the Institute.

1959-60

Prof. Chaudhuri joined as an Extension Officer of Fisheries in Calcutta. He evolved a methodology by which it was possible to import lakhs of Carp fish seed and spawn fish anywhere in India through a Syndicate.

1960-61

Prof. Chaudhuri went back to Cuttack and engaged in Fish Breeding Research and established Fish Breeding Research Institutes in various states.

1961

National Institute of Science awarded him Chandrakala Hora Memorial Gold Medal for his best success in Research.

1964

Food and Agricultural Organization (FAO) appointed him as an adviser on "Induced Breeding and on complicated processes in its application" at Taiwan. The Prof. did not join.

In the same year, he was invited by FAO for his leadership in "World Symposium on Warm Water Pond Fish Culture"

1969

Prof. Chaudhuri delivered three lectures on his success in various fields of fish research at a seminar organized by FAO/UNDP.

1969-70

Engaged as Adviser at Myanmar by FAO

1968

This year FAO organized a seminar at Russia, the main subject of which was: "Genetic Selection and Hybridization of cultivated fishes". Prof. Chaudhuri delivered lectures as a Chief speaker on Genetic selection and Hybridization of Carp all over Europe.

1972

FAO requested Prof. Chaudhuri to take over the job of an aquaculture expert in Bogota for three years. The Prof. was not released by the Govt. of India.

1974

FAO invited Prof. Chaudhuri to take the responsibility of milk fish breeding in Indonesia and thereafter International Development Research Centre (IDRC) appointed him as an Adviser of Madri Research Centre. The Prof. was not released.

1974

Prof. Chaudhuri worked for Indo-Pacific Fishery Council meeting on "Aquaculture and Environment".

1975

International Development and Research Centre (IDRC) in Canada appointed Prof. Chaudhuri as an Adviser of Carp Malaysia Project.

1976-79

SEAFDEC (Southeast Asia Fisheries Development Centre) appointed him as Regional Aquacultural Coordinator of the Aquaculture Department. At the same time he was appointed as the Deputy Director of Asian Institute of Aquaculture.

In the same year he got the invitation of 13th Pacific Science Congress. He was also appointed as the Project Coordinator and the Chief Speaker of IDRC's Rural Agricultural Project.

1976

Prof. Chaudhuri was also appointed as the Panel Member of FAO controlled Polyculture of Fin-Fish conference of Kiwoto in Japan. In the same year he was appointed as Regional Aquaculture Coordinator of SEAFDEC in the Philippines.

1978

IDRC of Canada appointed Prof. Chaudhuri as a Fishery Specialist of Malaysia and Sudan. He gave advice to the fish scientists of those two countries and provided help and advice in research.

In the same year Prof. Chaudhuri worked for the World Bank Mission dealing with eligibility and possibility of money investment.

The same year IDRC of Canada appointed him as an expert in fish breeding in Sarawak in Malaysia.

In the same year FAO requested to him to work in the different project of NACA project of SEAFDEC/UPV. He took the great responsibility of training aquaculturists, making the syllabus of MS degree and after working for one year he joined elsewhere.

1976

In hybridization work on *Labeo*, *Cirrhinus and Catla*, he created hybrids of five inter generic hybrids and hybrids of Rohu and Catla.

1984

Prof. Chaudhuri joined Laos as a Coordinator of FAO/UNDP Project.

1985-87

He was again appointed Senior Technical Adviser or SEAFDEC. Simultaneously he was given a great responsibility for training of FAO/UNDP/UPV trainees who were to be conferred Master of Science Degree in Agriculture.

1988-93

Prof. Chaudhuri joined as a Visiting Professor in the Institute of Biological Science in the University of Philippines.

A Chronological Account of Research Activities in India while in Government Service

1948

Junior/Senior Research

Riverine and Lacustrine Section in-charge.

Pond culture Section in-charge.

1949

Senior Research Assistant at Cuttack Worked on problem of heavy mortality of fish seed in Orissa.

1950-1953

Successfully developed technique to reduce seed mortality.

1954

Sent to USA by the Govt. of India on Point IV Pragramme. Obtained MS degree in Fisheries Management and other distinctions.

1955-56

At Central Inland Fisheries Research Station (CIFRS) succeeded in induced breeding of a carp minow *Esomus danricus* by hormone injection. Then succeeded in breeding a rivering Catfish *Pseudentropius atheriroides* by pituitary gland injection.

1957

On 10.07.1957 induced bred a member of difficult-to-spawn Asiatic Carp (*Cirrhinus reba*) by injection of Carp herones. Then it was followed with success in *Puntius sarana, Cirrhinus mrigala, Labeo rohita* and other carps. And a technique was subsequently developed and million of carp seeds were produced.

1957-59

Engaged in training of hundred of national and international fishery officers.

1958

Succeeded in hybridising carp by crossing the major and minor carps of the genera *Labeo, Cirrhinus* and *Catla* and obtained five interspecific hybrids and eight inter-generic hybrids and hybrid between rohu and common carp.

1959-60

Joined as fisheries Extension officer to advice and guide transport of millions of Carp Seeds to various states.

1960-61

Back to CIFRS at Cuttack. Continued research on induced breeding. Invited to various states to establish fish breeding centres.

1964

Took over charge of Cuttack Research Station as Research Planner, Research Guide and Administrator.

1964-68

On deputation of Burma under FAO/UNDP as Fisheries Adviser.

1968

Invited to USSR as a lecturer "On Genetic Selection and Hybridization of Cultivated fishes".

1971-75

As Senior Scientist and Officer-in-Charge of CIFRS, and Head of Fish Culture Division. Succeeded in producing 10 tons /ha/yr. by polyculture of Indian and Chinese Carps.

1976

Went to Voluntary Retirement.

1976-1993

Worked in Southeast Asia countries in various capacities.

A Summary
Extent of Hiralal's Activities
(Offer of assignments came from–FAO/UNDP, IDRC, ESCAP, WORLD BANK, SEARCA, SEAFDEC

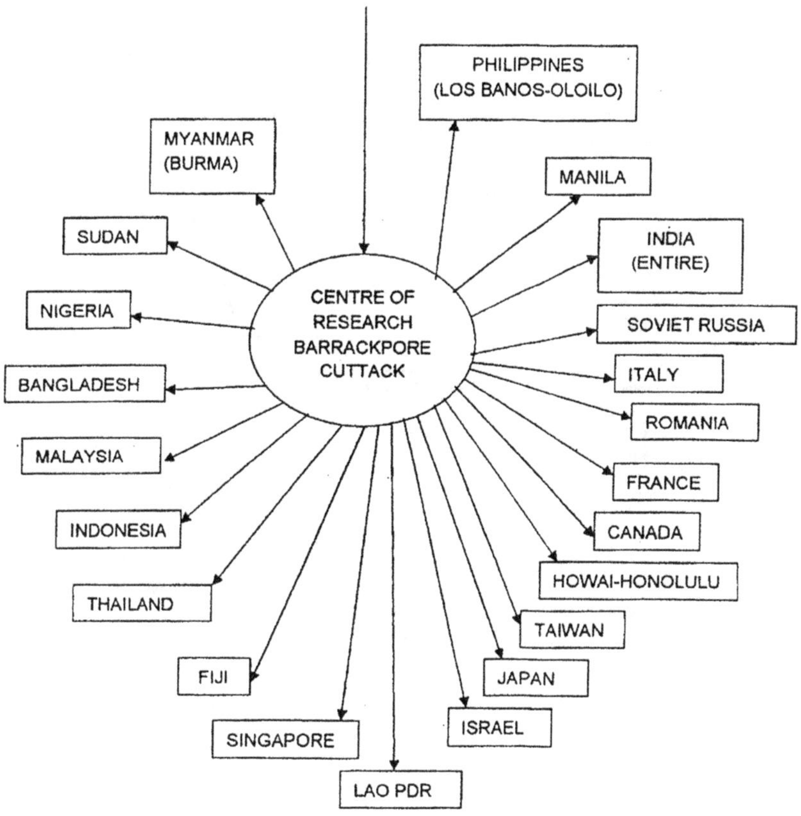

Hiralal's Sphere

of

Inland	Work/Activities	Abroad

Inland

1943-1974 Lecturer in Zoology

1948-1958 Junior Research Assistant in CIFRI

Induced Breeding success in 1957

1959-60 Extension Officer, Fisheries

1960-63 Fish breeding in-charge Class-I. Successful in induced breeding of *Mugil Cephalus (Bhangan)*

1964-66 In-Charge of Cuttack Station: (In 66 for 3 months in Burma)

1967-70 Abroad (Burma) Officiated as Director at Barrackpore on several occasions and On two occasions went abroad on Consultancy. In 1976 took Voluntary retirement.

Work/Activities

WORKED AS

- Researcher
- Research guide
- Research Planner
- Research Administrator

POSITION HELD BY HIRALAL

- Visiting Professor
- Technical Adviser
- Chief Technical Adviser
- Project Coordinator
- Special invitee for Lectures, Seminars. Work-shops.
- Fishery education.

Abroad

1945-55 Auburn University, USA for M.S. Degree. Fishery Adviser (FAO/UNDP) in Myanmar (1966 and 1967-70)

1976-79 Regional Aquaculture Coordinator (FAO/UNDP) SEAFDEC/SARCA and Deputy Director of Aquaculture. Successful in Induced breeding of Milk Fish

1979-84 Chief Technical Adviser & Coordinator Lao PDR. FAO/UNDP.

1985-88 Senior Technical Adviser at SEAFDEC in Philippines.

1988-93 Visiting Professor at UPLB, LOS BANOS in the Philippines

1994-2002 Various honours awarded in India and in California and Las Vegas in USA

AS ADVISER IN FIJI, SUDAN, NIGERIA, MALAYSIA, BANGLADESH, SINGAPORE, INDONESIA, TAIWAN, ISRAEL (MOSTLY)

Every Fibre of his bring was Aquaculture

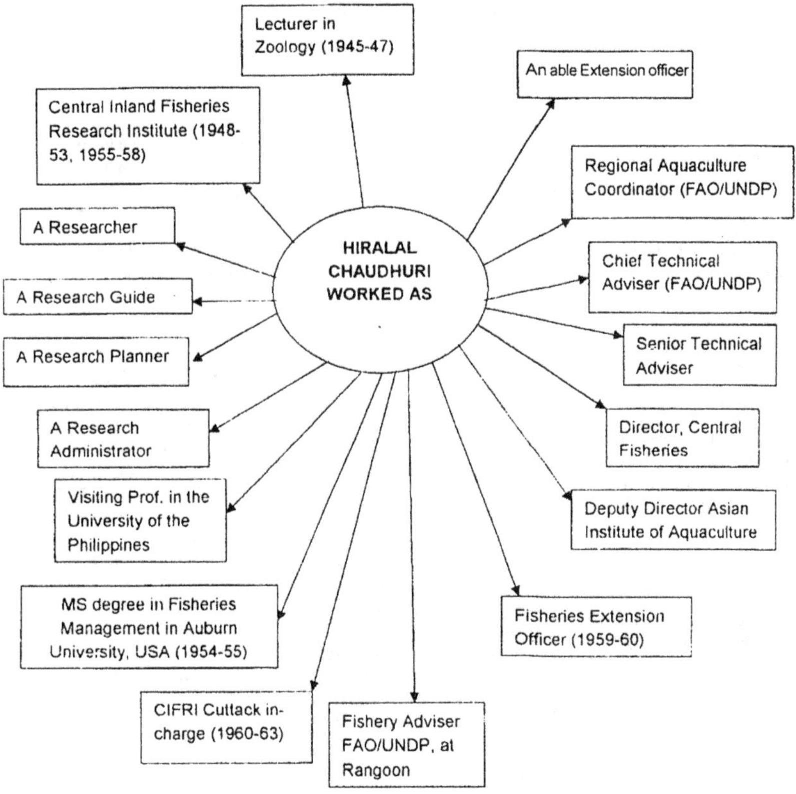

Prof. Chaudhuri's prestigious official posting
"Father of Induced Breeding"

[At Pacific Science Congress held at Vancouver in 1975. Dr. Kuronuma an eminent fishery Scientist of Japan welcomed Dr. Chaudhuri as the "Father of Induced Breeding". Since then this epithet has been an accepted fact.]

Technology break through of the Millennium when an Indian Crop was induced breed by hypophysation on July 10, 1957

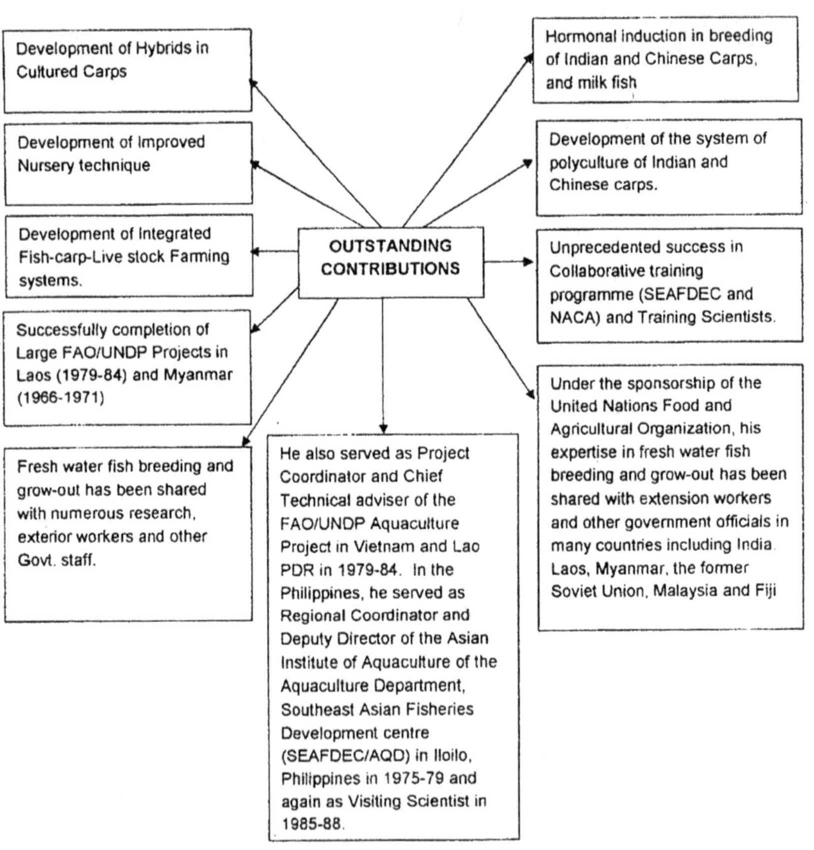

Chapter 4

Aquaculture in the Threshold of the 3rd Millennium: A Few Observations

A RESUME

Assessments and concepts of a few World famous aquaculturists about the progress of aquaculture in 80s and 90s of last century and its probable impacts in 21st Century have been considered very convenient and realistic to present their views in their own words.

The editor finds that most of these experts were unanimous in agreeing that the boom in aquaculture research successes which originated in the induced breeding of major carps and Chinese carps, massive fish-seed production, composite fish culture, hatchery and nursery management since, 1950s gained momentum over the years as a result of further research and introduction of new technologies. Prof. Chaudhuri's work it seems, was considered a pivotal source wherefrom aquaculture of 3rd millennium got a thrust.

15 such concepts including two of the Professor are presented in this chapter. "In the closing three decades of the 20th century the experts of Indian Fisheries reached its pinnacle with several of them entering the international arena mostly through FAO. In 52 years after independence India became a great fisheries nation observes 'Aquaculture Asia'.

This chapter records some stalwart aquaculturists concept of the state of aquaculture in the 3rd Millennium. All of them were of views that aquaculture has contributed substantially in solving protein food production in this country and that it has become fastest food production sector in the World. Aquaculture has expanded, diversified and technologically advanced.

On the Dawn of the New Millennium

A.   Professor Chaudhuri's Short Address at a National Workshop

To-day's National workshop is on "Bridging Gaps for Attaining self-sufficiency in Inland Fish Production at the Dawn of the New Millennium". If I have understood correctly, the main theme of discussion is about the gaps that exist between the demands and supply of inland fish and to find out ways and means as to how the gap could be bridged.

This reminds me of a conference held at FAO Head Quarters at Rome in 1970, where the theme of discussion was to envisage the future of World fisheries 30 years later at the beginning of the 21st Century. In that meeting, Dr. Roy Jackson, ADG (Fisheries) FAO, gave an interesting talk viewing the situation three decades ahead against the background of increasing world population and deteriorating aquatic environment. At the same time, he mentioned of industrial development, with concomitant alteration of the environment, pollution and consumption of resources, would be increasing exponentially in relation to population growth.

Thirty years have passed since then, and we are at the dawn of the millennium. Fish production, in the meantime had increased no doubt, but certainly not at the rate by which the population had increased and environmental conditions deteriorated. World population had almost doubled itself to around 6 billion and population of India had reached the billion mark.

While fish production could be increased manifold by application of advanced technologies, there is also the danger of polluting the environment. We are also aware that super-intensive culture could give extremely high production, very often resulting in bringing ecological dis-balance and adverse impact on the aquatic environment, mainly due to discharge of large quantities of metabolites. This has been a very big impediment towards achieving extra-ordinary high production. However, by proper application of Biotechnological approaches, such as genetic manipulations and Trans-genetic techniques etc., highly enhanced growth in fish could be obtained and production considerably increased. Principles of Biotechnology when applied to single-celled green alga *Chorella* or a cyanobacterium such as, *Spirulina* would not only increase phenomenal production of these fish feeds, but have also the potential to supplement human food as well.

Thus the situation we would be facing is that even if we are able to increase fish production extra-ordinarily, there would always be an upper limit while fixing our production target, beyond which we cannot go, if we have to protect our environment. Fish production, therefore, cannot cope up with the present rate of population growth and unless the latter is drastically cut, there would always be a large gap between expected supplies of fish and the potential demand, a situation like that of Tantalus' cup which can never be filled up.

I sincerely hope that in this workshop the participants would have useful deliberations on this problem, so that our future Fishery Scientists and Administrations, fish farmers and Fishermen communities, by their genuine collective efforts, will be able to at least narrow down this gap to a great extent.

B. Advances in Global Aquaculture

The editor has been able to retrive an abstract of an address made by Prof. Chaudhuri (details not known) in early 70s. This abstract confirms about his improved method of technology to increase fish food production. For Food-Fish Production he stresses of the Cage and Pen culture, Running water culture, Integrated fish-livestock crop farming system etc. He also deals with the application of hormones in fish seed production. All these works put aquaculture to a firm footing.

Extended Abstract

Aquaculture has been recognized in recent years, as one of the most important strategies to accelerate global fish production. The term 'Aquaculture' is defined as the planned or purposeful intervention in the production of fish and other aquatic organisms. Fish, as a group, is the final result of a complex biological cycle in aquatic ecosystem. Fish culture had been an age-old practice in Asia, especially, in China and India, but the average production had been generally low because of the traditional type of culture. After the Second World War, the aquaculture technology was gradually improved and more and more scientific methods were introduced. As a result, the fish production was increased from a meager 500-600 kg/ha/yr to 2-4 mt/ha/yr. However, during the last few decades, an all round development in the aquaculture industry had taken place. Consequently, the yield of fish has gone up considerably due to intensification and expansion of modern aquaculture techniques throughout the world. Such remarkable achievements could be possible mainly because of the application of the latest technologies in the two very important aspects of aquaculture, namely (*a*) Fish Seed Production and (*b*) Food-fish Production.

(*a*)   Fish Seed Production

The availability of adequate quantity of quality seed of the cultivated fish species, is an essential pre-requisite for obtaining large-scale success in the culture of any fish. Aquaculture industry in the world suffered tremendously due to the extreme dearth of seed for culture, since the majority of the major cultivable fishes either fair to reproduce in captivity, or do not attain fully maturity when reared in ponds of other enclosed waters. They generally breed in their natural habitats under optimum environmental conditions. The Asiatic carps (Indian and Chinese carps), cultured for centuries in ponds, belong to this category, as also the important euryhaline species such as, mullets, milkfish and sea-bass, the catadromous freshwater eels and majority of the commercially important marine fishes. For decades in the past, concerted efforts were made by fishery scientists to breed these fishes under artificial conditions. The successes was finally achieved in mid-1950's when members of Asiatic carps were induced to breed by injection of fish pituitary hormones. A dependable technique was subsequently developed to produce millions of seeds of these economic varieties of Indian and Chinese carps and induced fish breeding technique became a routine procedure in the various countries of China and Europe to produce quality seed to meet the demands of seed of these food-fishes for their culture. Following this success, in subsequent decades, it was possible to induce breed the popular grey mullets, milkfish, sea-bass, several catfishes and a few other species. More recently, a number of marine commercial species have been induced to spawn both natural and synthetic, as well as a few other hormones are progressively being used in recent years for inducing spawning in fishes. A few important hormones used in regular seed production programme, are mentioned below:

- ☆ HCG (Human Chorionic Gonadotropin obtained from human pregnancy urine) alone or in combination which FPE (Fish Pituitary Extract) gave varying success in breeding fishes and obtaining their seed.

- ☆ Luteinizing Hormone-Releasing Hormone (LH-RH) and its analogues. Here attempts had been made to intervene at the hypo thalamic hypophysis interface in the endocrinal chain instead of injecting exogenous gonadotropins to stimulate hypophysis directly.

- ☆ The mammalian (mGnRH) and /or piscine (sGnRH) gonadotropin-releasing hormone analogue combined with dopamine antagonists, is a further improvement over the previous one.

- ☆ Salmon gonadotropin-releasing hormone analogue (sGnRH-A) in combination with domperidone, known as 'Linpe Method' is largely used in China in the large-scale seed production of Chinese carps. 'Ovaprim' its trade name, prepared by Syndel Lab., Canada, is used for induced breeding in several countries.

However, the poor fish farming of majority of the developing countries of the world still prefer to use the whole pituitary gland extract mainly because of the lower cost, simpler technique and ready availability of the hormones.

The induced breeding technique has also opened a new line of research on the genetic selection and hybridization in these fishes for production superior strains of fishes.

(b) Food-Fish Production

With the expansion of cultural areas and intensification of culture, the overall production of fish through aquaculture had increased considerably. Besides, by the introduction of scientific cultural methods technology including improved management practices, such as, stock manipulation, pond fertilization, supplementary feeding, disease control etc., the fish production could be increased from a meagre 500-600 kg/hr/yr to about 4-5 mt/ha/yr.

Further, with the introduction of exotic species of superior qualities such as the faster growing Chinese bighead, silver carp, grass carp etc., along with Indian major carps of an elaborate system of culture is being practiced, known as polyculture or mixed culture of a number of fast growing compatible species of fishes having different feeding habits and different Seeding zones. By polyculture of a large number of species, it is possible that the different varieties of food that are produced and are available in the pond environment, could be fully utilized and converted to fish flesh, by mixed-fish farming the production could be increased from 8-10 tons to as high as 20-25 mt/ha/yr.

Also, introduction of a faster growing species of tilapia (*Oreochromis niloticus*) in many countries of Asia and the Pacific, the fish production had been increased tremendously. Further by genetical research and application intensive farming technology, a red hybrid strain has been produced which has been reported to have given a production as high as from 600-800 mt./ha/yr.

The fish production could further be enhanced considerably by following certain Non-conventional methods of culture, such as, Cage and Pen Culture, Integrated fish–livestock-crop farming systems and Recycling of Animal Wastes and Running water fish culture.

Integrated Fish-livestock-crop Farming Systems

This is a very high yielding farming system where aquaculture is combined with simultaneous raising of livestock and vegetable/crop farming. In this system, there is mutual benefit among the different commodities and the wastes and by-products of one is utilized by another and practically nothing is wasted. Fishes get the maximum benefits since they utilize the wastes and by-products of livestocks and crops, thereby saving expenditures on feeds and fertilizers. Recycling of animal wastes is very very important since it converts the animal wastes to fish flesh by fertilizing the pond waters which otherwise would have been a nuisance and a health hazard. By simple recycling of wastes, 5-10 mt/ha/yr fish can be produced without using inorganic fertilizers or providing any supplementary feeding.

Cage and Pen Culture

Cage and pen culture, although an age-old practice in a few countries of South-east Asia, the modernization of the cultural techniques have in recent years given very high yields in fresh, brackishwater as well as through mariculture. In Japan, millions of metric tones of yellow-tail and sunbreams are produced every year in marine cages.

Running Water Fish Culture

The principle behind running water fish culture is that the flowing water constantly removes the nitrogenous and other metabolites from the culture waters, thereby eliminating these toxic products which when accumulated inhibit fish growth. Running water culture can give unbelievable production, such as, over 4000mt/ha/yr (estimated) of common carps in Japan and over 2000 mt/ha/yr of rainbow trouts in Idaho, USA. Such high production is possible because of intensive stocking, continuous elimination of metabolites and feeding ad libitum in oxygen-rich flowing water.

To sum up, it may be confidently said that recent advances in aquaculture practices have shown that the industry has reached such a stage that is capable of providing a substantial percentage of animal protein to the diet of protein deficient poorer masses of the third world countries.

C. Aquaculture and Dawn of New Millennium

Dr. Monoranjan Sinha, Director CICF writes (Fishing Chimes vol. 19, Nov. 1999) that induced breeding and seed rearing have revolutionized the aquaculture practices. Excerpts from his presentation which speaks of Prof. Chaudhuri's earlier successful works are as follows:-

Technologies of induced breeding, seed rearing have virtually revolutionized the aquaculture practices in the State. The three-tier system of carp seed raising and culture have provided for specific production measures for the stages of fry and fingerlings for stocking in grow-out ponds.

Seed Production Systems

West Bengal, the pioneer State in carp seed production, alone contributes to about 75 per cent of the total seed production of the country. This has been possible owing to the impact of intensive training and demonstration efforts of Central Inland Fisheries Research Institute, Directorate of Fisheries and Non-Government Organizations, resulting in the establishment of large a number of Glass jar and Eco-hatcheries throughout the State. This achievement has not only salvaged the State from long dependence on collection of natural riverines source of fish seed but also helped to produce about 45 per cent surplus to cater to the needs of other States of country. The carp seed industry in the State is dependent on Riverine seed collection, bundh breeding and induced breeding techniques.

The Asia Regional Planning works of September 1999 agreed on a vision for aquaculture and a mission for the region that "Aquaculture shall be a major provider of food and will be applied to reducing social disparities and inequalities and the mission will be to:

☆ Fulfill the aspirations of the sectors of the populace through a people-oriented approach focusing on aquaculture for development.

☆ Develop aquaculture in a responsible manner in harmony with the environment.

☆ Transform the emphasis of aquaculture form a resource-dependent towards a knowledge-based activity; and

☆ Continue to pursue the shared objectives of the governments through regional cooperation and strengthen.

Prof. Chaudhuri's 45 years of service for Aquaculture satisfies most of the 17 items of strategy elements in the Bangkok declaration of February 2000. His association with NACA was very close and extremely productive.

The observations made by the editor of "Aquaculture Asia" in Oct-Dec. 2000 issue are presented in part to register that Prof. Chaudhuri work involved most of the 17 items of strategy elements made in Bangkok declaration. The observations are partly presented.

For world aquaculture, the Bangkok Declaration and Strategy for Aquaculture Development Beyond 2000 on Aquaculture in the Third Millennium held in February 2000, provides a blueprint for government action for aquaculture development, as outlined by the 17 strategy elements identified by the Conference which are investing in people through education and training, investing in research and development, improving information flow and communication, improving food security and alleviating poverty, improving environmental sustainability, integrating aquaculture into rural development, investing in aquaculture development, strengthening institutional support, applying innovations in aquaculture, improving culture-based fisheries and enhancements, managing aquatic animal health, improving nutrition, applying genetics, applying biotechnology, improving food quality and safety; promoting market development and trade, and supporting strong regional and inter-regional cooperation.

The Bangkok Declaration urges that aquaculture be developed to its full potential so it can make a net contribution to food security, economic growth, trade and improved living standards; that it be a component of development contributing to sustainable livelihoods of the poor; that policies promote farming practices that are environmentally and socially responsible; that states, their private sector and other legitimate stakeholders cooperate to promote the responsible growth of aquaculture; and that regional and inter-regional cooperation be strengthened to raise the effectiveness of aquaculture development efforts.

The directional guides from these regional and global events suggest a 7–pronged thrust to Asian aquaculture development during the first decade of this new millennium, as follows:

1. A basic shift in emphasis from aquaculture development to aquaculture for attaining social development objectives;

2. Further emphasis on environmental sustainability;

3. Harnessing and integrating science-based and indigenous knowledge to improve aquaculture technology and systems and make these relevant to people's needs and circumstances;

4. Increasing reliance on information technology to develop and deliver environmentally sustainable innovations;

5. A strategic shift in the networking structure from institutional to people-centered networking;

6. An increasing reliance on technical cooperation among states; and

7. A greater participation and more active involvement in inter-regional co-operative actions.

An essential factor to underpin these thrusts will be quality and relevant education at all levels delivered cost-effectively, for which a regional collaborative effort has been initiated.

Going Global

Global cooperation, in most likelihood, will be an obligatory direction for Asian aquaculture. Ditto for the intergovernmental NACA Organization whose membership currently stands at 14 states in the region with five more active participants in Asia and one in West Asia. (Seven of these are among the world's top 10 aquaculture producers). During and in the aftermath of the Bangkok Conference, the NACA mechanism was being looked at as a model for the other developing regions, particularly Africa and Latin America. Veteran aquaculture development advisers in these two regions in fact were using the word "revisit" to refer to the fact that a global aquaculture development coordination project had established regional networks in Africa, Asia, and Latin America, with the networks attaining varying states of development (Pillay, 2000).

FAO, the day after the Aquaculture Conference, convened a team of global experts to look into an appropriate way to carry out global cooperation in aquaculture development. The consultation suggested creating a subcommittee on aquaculture under the Committee on Fisheries; priority activities can be identified and plans and strategies developed from there. For starters, six key issues the subcommittee should address were identified: role of aquaculture in food security, economic development and poverty alleviation; consumer issues as in food safety, quality and certification; human resource development, research and extension; environmental aspects of aquaculture development; institutional capacity building and policy development; and statistics, data and information management (NACA Newsletter Jan-Jul 2000). The first half of the 20th Century although was sporadically significant because consequently pale compared to the distinctive and fast forward stride in the second half of the century.

The editor doesn't intend to present the full history in support of the aforesaid observation except that induced breeding, organized pond culture, development of fishery in reservoir, beels and in other fields.

An observation of Fishing Chimes is presented below:

"By 1997-98 total fish production went up to a level of 54 lakh (29.50 lakh temarine and 24.38 lakh inland) from 0.52 lakh in 1950-51 (0.34 lakh marine and 0.18 lakh inland). Several Fishery Research Institutes (CMFRI, CIFRI, CIFT, CIBA, NBFGR, NRCWF), a Fishery survey outfit under various successive names but ultimately designated as Fishery Survey of India, Integrated Fisheries Project as successors to Indo-Norwegian Project, CICEF (Central Institute for Coastal Engineering for Fishery), CIFNET (Central Institute for Fisheries Nautical Engineering Training), Several Fisheries Training Centres, and MPEDA (Marine Products Export Development Authority) were set up. The National Institute of Oceanography came into being up in late sixties".

While CIFRI gave to the nation the induced breeding and pond culture technology (eventually inherited by its offshoot, CIFA). This was followed by the setting up of a no. of jar and Chinese hatcheries. The Indo-Norwegian project presented to the nation designs and construction and operational aspects of mechanized fishing boats, later supplemented by other designs and motorized boats by CIFT. BOBP, started in early sixties of 20th Century introduced several innovative technologies relating to design and operation of fishing boats, nets, smoking of fish, fish transportation etc. and in respect of socio-economics of fishermen.

Fishing Chimes thereafter mentions various developments in the design, construction and operational aspects of motorized boats, especially in coastal sector. It also speaks of Prof. Chaudhuri's epoc-making contribution to aquaculture which reads.

"A further remarkable achievement of the second part of the century was in the sector of culture fish production. It reached new heights, supported by self sufficiency in seed production. Induced fish breeding technology through hypophysation and through administration of synthetic hormones was introduced and the techniques thereof spread over the entire length and breadth of the country. In the case of shrimps, the technology of inducing them to breed through eyestalk ablation was developed and became cultural. Efforts were on for applying the technology to mud crabs. In the same way, a few trout hatcheries in the high attitude areas in Himachal, UP and J&K were established in addition to those set up by the British in the first half of the century. The last few decades of the century saw the setting up of a large member of major carp hatcheries and establishment of a several shrimp hatcheries and the setting up of several shrimp farms dotting most of the coastline. The nation reached take-off stage in respect of genetic up-gradation of cultivable fish species in the last decade of the century. So was more or less the case in respect of reservoir fishery development. An awareness developed that (a) reservoir potential for fisheries development should be harnessed, (b) Underground saline water reserves in north-western states should be utilized for augmenting fish production, (c) running water fish culture and wetland fishery development should be given a fillip with an export orientation".

In the closing three decades of the century Indian fisheries expertise reached its pinnacle, with several experts entering the international arena, mostly through FAO. In a short history of a little over 52 years after independence, India became a great fisheries nation, respected the world over for its achievements in the sector. While there is much more to be accomplished, it is great to see the nation on a firm course to achieve new heights of endeavor in the fisheries sector for the benefit of the nation and the people.

Indian Freshwater Aquaculture in the Coming Millennium

The discoveries of Prof. Chaudhuri in the second half last century and his vision of next millennium gets a firm support in having a realistic assessment in an article presented by Dr. S. Ayyapan (Source: ISFP, Nov. 1999, Vol. I No 4) which this editor finds very relevant and that reflects Prof. Chaudhuri's useful findings are as follows:

Indian freshwater aquaculture has evolved itself from the stage of a domestic activity in the Eastern States of West Bengal and Orissa to that of an industry over the years, with entrepreneurs in the States like Andhra Pradesh, Punjab and Haryana undertaking the practices in a big way. The sector with an estimated production of over 1.9 million metric tonnes of fish is contributing to the extent of 34 per cent of the fish basket of the country. Being highly compatible with other farming activities, with flexible scales of investments, the sector has been showing an impressive annual growth rate of over 6 per cent. Technological inputs, financial investments and entrepreneurial interest have enabled enhancement of productivity levels from 600-800 kg ha-1 yr-1 to a national average of more than 2000 kg ha-1 yr1, with achievements of 8–10 t ha-1 yr-1 in several parts of the country. Intensification of carp culture systems that form the mainstay of Indian freshwater aquaculture. Diversification of culture practices in term of catfishes and freshwater prawns as also pearl culture, integrated fish farming recycling into host of organic wastes into fish culture systems, generic improvement of fish species, incorporation of supplementary feeds in the management practices, fish health management are some of the aspects that have enabled the growth rates.

Recent accomplishments in researches in freshwater aquaculture in the country include prolonged and multiple breeding of carps (as many as four times) from April to September for seed availability throughout the year, carp milt cryopreservation for brood stock enhancement, high density seed rearing with spawn at 50 million per ha and fry at 5 million per ha, production rates 10 and 15 tonnes ha-1 yr-1 under intensive carp culture, breding of *Clarias batrachus* (magur) and *Heteropneustes fossilis* (singhi) under controlled conditions and development of hatchery designs, breeding and culture of Wallago attu and other major catfishes, brood stock development and hatchery management for giant freshwater prawn, *Macrobrachium rosenbergii* and the riverine prawn *M. malcomsonii*, production of freshwater pearl through implantation of nuclei in freshwater mussel, *Lamellidens* sp. Aquatic biofertilization measures using Alolla at 40 t ha-1 yr-1 and application of processed organic material like biogas slurry as manurial inputs, treatment domestic sewage through algal-duckweed-fish culture, diploid gynogens and polyploid carps like triploid grass carp, production of improved rohu through selection, formulation of prophylactics and therapeutics against fish diseases, design and development of cage culture, recirculatory and flow-thru' systems.

Considering the global scenario of R and D in freshwater. aquaculture and the resources and needs of the country in the coming millennium, the following areas of research are identified as important areas of work in the coming years. Intensive carp culture practices with sustainable production targets of 25-30 t ha.1 yr-1; evaluation of new carp and catfish species as candidates for culture; standardization of techniques of cage culture and pen culture for ranching large water sheets; aquaculture in water harvest structures and as an integral component of watershed management; diversification of aquaculture practices along with freshwater prawn and pearl and new candidate species like crabs, tortoise, algae and aquatic plants like Spirulina; standardization of indigenous GnRH formulations for induced breeding of fish; cryopreservation of carp and catfish gametes and embryos and multiple breeding; self-contained breeding and hatchery technology for freshwater prawns; large scale cultured pearl production with post-harvest processing; composite fertilizer management; use of microbially processed organic inputs and biofilteration; aquatic environment-produce interactions with EIA studies; identification and cloning of single genes and confer commercial benefits in cultured fish/shellfish species through recombinant DNA technology; genetic engineering with reference to growth and disease resistance in fish/shellfish species; strain improvement through selection; feed dispensing methodologies; serological studies to develop indices of health conditions; immunological studies and drug sensitivity evaluation with respect to selected microbial infections and post-harvest technology for fresh water fish /shellfish species.

Considering the prevalent production levels in ponds and tanks in different States of the country with due reference to the agro-climatic conditions, fish species, technology adoption levels, demand patterns, export potentials and predicted growth in freshwater aquaculture sector in different States, a strategic plan, Operation Aqua Gold *(Matsyavardhan)* is proposed, for doubling the fish production from the freshwater aquaculture sector. It is estimated that 23,58,634 ha of ponds and tanks are available for culture practices in the country. Of the total inland fish production of 22,42,170 t, an estimated 15,12,000 t come from the freshwater aquaculture systems. Considering the prevalent mean national pond productivity of 183 kg ha-1 yr-1, it can be deduced that about 8,26,230 ha are presently under fish culture.

Under operation Aqua-Gold (Matsyavardhan), it is proposed to increase the aquaculture area coverage to 11,99,500 ha, with a mean productivity of 2,762 kg ha-1 yr-1 that would yield 33,12,800 t, amounting to doubling of the present fish production. For achieving increased fish production from freshwater aquaculture by 2.19 times, it is necessary to bring an additional area of ponds and tanks to the extent of 3,73,270 ha and increase the pond productivity by 932kg ha-1 yr-1, amounting to an increment of 45.2 per cent and 50.9 per cent in the two approaches respectively. The strategy considers both horizontal and vertical expansion for achieving the desired results of doubling fish production from the sector. While the present carp seed production of 15007 million fry are being used for both culture and culture-based capture fisheries, the latter in terms of stocking the reservoirs, the projected requirements of seed for culture purpose are nearly of the same order. For producing the required quantity of 45000 million spawn for the purpose, about 900 t of brood stock is required to be maintained over an area of 80000 ha, that is 6.7 per cent of the projected culture area in the country. While most of the freshwater aquaculture is based on the organic fertilization presently in the country, use of supplementary feeds for increasing fish productivity with a view to realize the full potentials of the waters is emphasized. An estimated 52 lakh tonnes of fish feed would be required for achieving fish production levels of 2t ha-1 yr-1 and above from about 73 per cent of waters in the projection.

E.   Freshwater Culture
Drs. H.P.C Shetty and G.P.S. Rao (World Aquaculture, March 1996) have Observed the Following:

Freshwater aquaculture has been practiced in India since ancient times. However, until the middle of the present century, aquaculture has been a backward activity found mainly in the northeastern states. Due to progress made over the last three decades, there is now high export potential for Indian fresh water aquaculture products.

Resources available for freshwater aquaculture are considerable in terms of water and species are suitable for culture. According to the Central Institute of Freshwater Aquaculture, the estimated available water area for aquaculture or culture based capture fisheries is 2.25 million hectares of ponds (including stock watering ponds), 1.3 million hectares of extensive water bodies in low lying areas (known locally as beels and jheels) and derelict water bodies, 2.09 million ha of lakes and reservoirs, 2.3 million hectares of paddy fields (Portions of which are suitable for aquaculture), and 120 thousand kilometers of canals and channels.

There are about 75 species of carp, catfish, clupeid fishes, mullet, murrels, perch, freshwater shrimp, and molluscs suitable for culture in India, but only about two dozen species are currently being reared by commercial culturists.

India is basically a carp producing nation. The Indian major carps catla rohu, and mrigal and the exotic common carp, grass carp, and silver carp are the most widely grown species. Some of the medium carps, minor carps, and catfishes are in great demand in certain regions. However, there is no commercial culture of the latter species at this time.

Carp Culture

Carp culture systems are highly varied, depending on the availability of fry or finger lines for stocking and investment capital. The simplest form of extensive fish culture is practiced in water bodies from 5 to 25 ha that are leased to growers by government agencies. The fish grow on natural food and are harvested after 9 to 12 months of rearing. Fish production under this system varies from 500 to 1600 kg/ha/yr. depending on pond fertility. Encouraged by the profitability of scientific fish culture, the farmers of Andra Pradesh, especially in the coastal districts of West Godavari and Krishna surrounding the Kolleru lake area, have converted more than 40 000 ha of paddy fields into fish ponds where the level of culture is semi-intensive. The species being reared are the Indian major carps (catla, rohu, and mrigal). Ponds are provided with both organic manures and inorganic fertilizers. In addition, rice bran is provided as supplemental feed. Farmers obtain an average of 3000 to 5000 kg/ha/yr using this method. Some farmers stock fish with initial sizes of 100 to 150 g and obtain yields of 8000 to 10,000 kg/ha/yr. Ponds range in size from 1 to 50 ha.

F. George Chamberlain and Herald Rosenthal Observe (World Aquaculture, March 1995)

"In the last decade, aquaculture has been the only growth sector within fisheries and the prospects for continued growth appear excellent. Global per capita seafood consumption has been rising steadily since 1969, but landings from the capture fisheries reached a plateau in 1989, leaving aquaculture as the primary source of seafood production to meet this increasing demand.

A substantial portion of the global increase in aquaculture production has come from coastal environments, but as the human population grows and expands its involvement in the coastal zones, there will be increasing pressure to share the coastal resources among multiple users. In this environment some of our existing aquaculture practices will not be sustainable in their present form, but those that area designed to accommodate multiple resource use could grow rapidly. Examples range from the traditional farming systems in Southeast Asia, which benefit the community at large as well as the aquaculturists themselves, to modern high-tech recirculation systems."

G. Aquaculture Conferences
K.C. Lucas, Assistant Director-General (Fisheries) FAO Observes

"To many countries aquaculture in something of a novelty. To others it consist of traditional products which have a history of many countries some find it a source of supply of animal protein food, other considers it to raise high-priced aquatic species for profit".

"Some say it involves improvement of aquatic environment through farming and intensive culture. It includes a great variety of plants and animals."

The Kyoto conference considered the following, which brought together in one forum a broad spectrum of scientists, techniques administrators, entrepreneurs and financiers.

H. Aquaculture Growth Between 1976 and 2000

The editor of "Aquaculture Asia" records that past quarter of a century (1976–2000) since the FAO International Technical Conference on aquaculture held in Kyoto in 1976, aquaculture has evolved from a traditional to science based activity to stride into a sector that is producing more nutritious food, providing better livelihoods for farm families and contributing more to national

Ramifications of Aquaculture

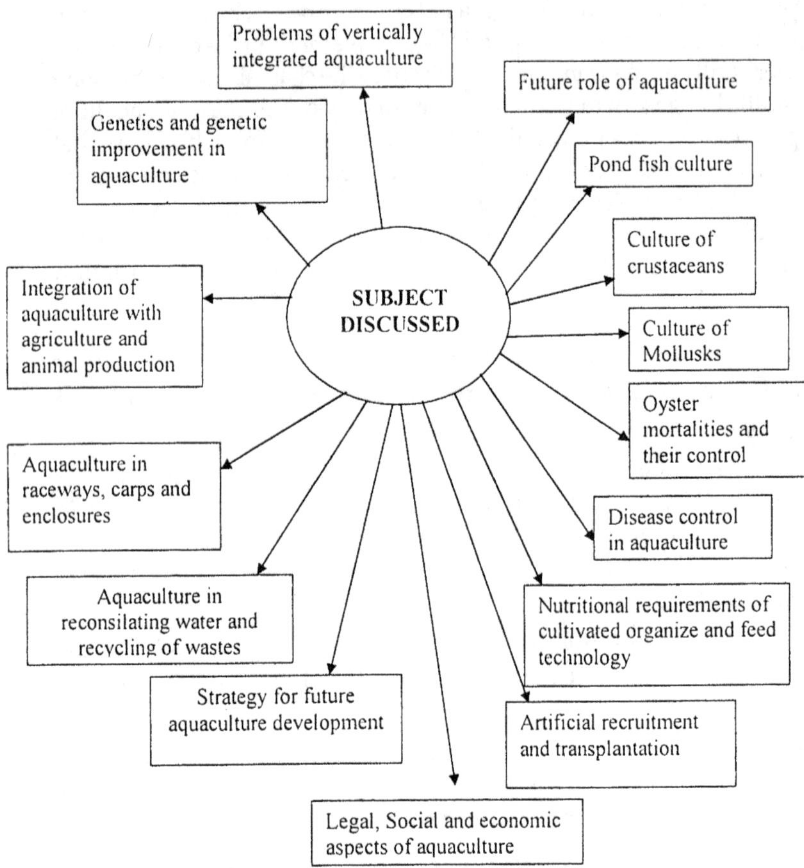

economies. Kyoto Declaration and strategy had served as the guide for global aquaculture development the past 20 years.

Since 1976 the fish production has an 8 fold increase. This was largely attributed to the increase in government planning and development inputs, which elevated the status of aquaculture to a level that is close to that of capture fisheries (NACA 1999).

The Asian Aquaculture development in the new millennium, the Asian Regional planning workshop of September 1999 agreed on a vision for aquaculture and the mission for the region that "Aquaculture shall be a major provider of food and will be applied to reducing social disparities and inequities", and the mission will be to:

- ☆ Fulfill aspirations of all sectors of the populace through a people–oriented approach focusing on aquaculture for development;
- ☆ Develop aquaculture in a responsible manner in harmony with the environment;
- ☆ Transform the emphasis of aquaculture from a resource dependent towards a knowledge–based activity; and
- ☆ Continue to pursue the shared objectives of the governments through regional cooperation.

I. Aquaculture developments Beyond 2000: Global Prospects

Over the past three decades, aquaculture has developed to become the fastest growing food producing sector in the world. Aquaculture has expanded, diversified, intensified, and technologically advanced, and its contribution to aquatic food production has increased significantly. Aquaculture is highly diverse and consists of a broad spectrum of systems, practices and operations, ranging from backyard small, household pond systems to large-scale, highly intensive practices. A large proportion of aquaculture production comes from small-scale producers in developing and low income food deficit countries (LIFDCs) and the sector contributes to food security, poverty alleviation and social well-being in many countries. Contributions of aquaculture to trade both local and international have increased over the past decades, and its share in the generation of income and employment for economic development has increased in many countries.

[Source: Aquaculture Asia 2000]

J.   Aquaculture and Sustainability

John Bandach writes (World Aquaculture, March 1995) about aquaculture vis-a-vis increasing demand for protein. Only shall excerpts of his article presented are as follows:

The world fisheries are not likely to satisfy the increasing demand for animal protein of aquatic provenience and aquaculture will clearly be called upon to make up the shortfalls. Water farming, with all its permutations, has grown by leaps and bounds in the last few decades, but not without increasing concern for the environment. Competition for water as well as for suitable sites, particularly on the coasts, is fueling increasing conflict. Broadly speaking, there is a growing challenge to make aquaculture compatible with other demands on water, land and near it more sustainable.

All in all, sustainable development is a broad goal for many processes and endeavors as desirable, of course, as peace, health and the like. The difficulty is in the details of how to attain it; a conundrum that applies to aquaculture as much as to any economic activity. These details differ with different modes of aquaculture from growing algae for colloid production to the rearing of carnivorous animals at high densities. So important are these differences with regard to energy and material flows through the various systems that generalities become meaningless and perhaps the one demand that can be made of aquaculture is that of the classical guidance to the physician, "First, do no harm." Harm to the environment that is, the immediate environment in which we function and that left for the coming generations is a paramount issue if aquaculture is going to help to "feed the world," as some have forecast.

To this observer of developments in aquaculture over several decades, it seems that the demand for economic sustainability is becoming increasingly synoymous with ecological sustainability, especially as demand for products and pressure by humankind on all aquatic environments increases. The better this connection between economics and environmental consequences is made, the more hopeful becomes the future of sustainable aquaculture.

K. Food Situation in 21st Century

Corner Bailey in his presentations "Aquaculture and Basic human needs" (*Source*: World Aquaculture, September 1997) reviews food situation in 21st Century.

Some excerpts from his article are presented below. The conception of Prof. Chaudhuri was that sea food aquaculture will play a vital role in solving food problem of the country which Bailey's observation up holds.

Food

Readers of "World Aquaculture" are aware that most of the important wild fish stocks are fully exploited, and in many cases overexploited. Demand for seafood is likely to expand as a result of growth in human populations and rising incomes in non-industrialized nations of the world. The current world population is rapidly approaching 6 billion and is projected to exceed 8 billion by the year 2020, reaching perhaps 10 billion by 2050 before stabilizing at nearly 12 billion by the end of the 21st century. Most of this growth will occur in developing nations, especially in Asia, where rising incomes are likely to spur demand for food generally, and high quality animal protein in particular. Much of this protein demand will be met by aquaculturists, especially in Asia, where over half the human population resides, where aquaculture is an established industry, and where fish, crustaceans, and shellfish already are widely consumed. In many developing nations, fish is the only affordable source of animal protein, a situation likely to continue in the foreseeable future.

The role aquaculture will play in meeting human needs for high quality protein will principally depend on what kinds of Aquacultural production systems are developed and implemented in the future. The crux of the issue is whether the emphasis will be production of high-value luxury commodities for export to wealthy consumers of industrialized nations, or Aquacultural development to meet the nutritional needs of local populations. There is nothing inherently wrong with producing for an export market which provides high income for producers and foreign exchange earnings for nations that need hard currency. The issue is, rather, the relative emphasis placed on one or the other type of development.

The current reality is that high prices for shrimp and other luxury commodities attract investment of human and, financial resources from both private and public sectors. Many producers strive to maximize yields per unit area in the belief that this strategy will provide the highest and most rapid return on investment, even though this goal is not always achieved and both ecological and financial risks are dramatically increased in the process. Recent research on shrimp aquaculture in Indonesia demonstrates that semi intensive systems have higher rates of return, involve lower risk, and generate more employment than intensive systems. Nonetheless, intensive systems remain attractive because, while rates of return are lower, absolute returns may be higher, at least in the short run.

Market forces, governmental policies, and the nature of scientific research all push us in a direction that is in conflict with efforts to meet human needs for food, viable social and economic systems, and healthy ecosystems.

L. Water Crisis in South and South-East Asia

Is inland aquaculture in crisis

Does this indicate a bleak future for development of freshwater aquaculture in South and South-East Asia?

With the largest population density in the world and widespread poverty, excepting the island areas the region of South Asia covers about 3.3 percent of the terrestrial area and receives 6.8 per cent of the annual replanishable amount of water of the world with this land and water resources, the region supports 21 per cent of the world population. The challenges in water systems management are rooted in the common objective of all the countries– poverty alleviation and sustainable development. The growing and conflicting demands from food security, commercial farming, domestic water needs of the rapidly increasing urban population, hydropower projects, industrial demands, damage due to rampant population, etc. need to be rapidly solved with informed political wisdom.

[*Source: J. Bandopadhyay in Everyman's Science,*
Vol. XLI, 05.10.1997.]

M. Biotechnology and Aquaculture in next Millennium

Prof. Chaudhuri's concept on the application of biotechnology and its application to aquaculture was presented by this editor in his book on the professors technical contribution. So the readers may refer to that presentation.

He was aware of the techniques of molecular biology, genetic manipulation methods, and the use of growth hormones that have been playing a beneficial role in aquaculture. He had the concept of development of vaccines through biotechnology to contain disease.

Role of Biotechnology

In the near future a considerable advancement in the use of biotechnology in aquaculture is expected for increase in production in terms of a quality and value addition.

Science of nutrition has not played its due role as yet in Indian aquaculture industry. A promising approach would be to improve the quality and utilization of various agricultural and industrial by products for efficiency. Several foodstuffs which have potential value as fish feed ingredients contain toxic or anti-nutritional factors affecting nutrient utilization. Biotechnology supported by traditional plant breeding practices may lead to development of new varieties or cultivars free from or with low content of such toxicants.

Identification of feeding stimuli/chemical signals is necessary to obtain the full sequence of feeding behavior from initial recognition of feeding "strike" followed by ingestion of food particles. It may be possible now genetically to design or engineer the bacterial strain with characteristics of having better cellulose degradation capacity. Thus the use of "bypas" nutrients would help improving the utilization of nutrients from agro-industrial by products.

Microbial processing of the livestock manure sure as anaerobic digestion in biogas plants reduces the in pond oxygen demand and improves fertilization quality of the inputs. Increasing the nitrogen fixation levels in ponds through bio fertilization with blue green algae, Azolla and genetic manipulation and bacteria to improve their nitrogen fixing capacity are other important areas where biotechnology could be of immense significance in aquaculture development and increasing fish production.

Mammalian growth hormones administered through feed or injection enhances fish growth, Gonadotropic hormones are being applied for spawning of carps and catfishes. 17 L-hydrozy-progesterone is a well known inducing agent for maturation and ovulation. LHRH and its analogues are effective in inducing gonadotropin release. Dopamine antagonists such as pimozide and domperidone are known to potentiate the effects of LHRH. Hormonal manipulation of sex is practiced to control unwanted reproduction in prolific breeders such as Tilapia and common carp. Manipulated monosex population improves nutritive value of fish flesh and growth. Implantation of hormonal pellets in fish body helps in bringing about maturation and stimulates advance or multiple spawning in fish. Successful protocols for cryopreservation of milt have been developed for several fishes. Cryopreservation of zygotes and embryos would facilitate large scale fish seed production programme.

Use of allomones and pheromones in large scale induced breeding could be of great significance. Role of pheromones in alarm and social behaviour, species and sex recognition, sex behabiour and territorial and space recognition would have to be studied under biotechnology programme.

Fish stock can be improved for breeding and culture through chromosome set manipulation by way of gynogenesis / androgenesis and polyploidy. Sterile triploids are produced to control over population and improving growth rates in the fish stock.

Development of vaccines through biotechnogy would be to bring sophistication in fish diseases management. Vaccines against virbiosis or furunculosis are available but these are for limited fish species or bacterial pathogens. There is an increasing need to use biotechnology for developing vaccines against other bacterial, viral parasities diseases of several commercially important culturable fish species. Investigations on immunity to the virus infections of fish need to be encouraged.

Development of biotechnology research in fishery sector would urgently require upgradation of laboratory facilities and training of specific manpower by Indian Council of Agricultural Research.

The current breakthrough in various technologies in non-food fisheries such as freshwater and marigepearl culture, product development such as citin and chitosan etc., need to be further studies for even more efficient and large scale production technologies. Attention is also drawn towards genetic improvement of fish stock through selection and genetic engineering, ex-situ conservation methodologies, gene banking both with live and gamete level approaches are certain essential trends for near future reach. There has been obvious neglect of studies on social impact of fisheries and aquaculture development. The need is to develop parameters to undertake Social Impact Assessment (SIA).

It may be appreciated that the land and water resources in the country are not available exclusively for aquaculture as there is excessive pressure on the resources from several other sectors. Any further growth in aquaculture for quantum jump would have to be vertical and technology based. The transfer of technology in the country has had its own casualties. Even today, the proven technology available in various fields of fisheries and aquaculture find its manifestation in the field at as low level as 30 per cent. The gap between the results of research and those in the farmers fields have to be narrowed.

[Source: Dr. B. Trinadha Babu and Y. P. Rao in
Fishing Chimes, August 2000]

N. "Aquaculture Asia": Observes (Oct.–Nov. 2000 Issue)

"A quick review of achievements in Asian aquaculture since 1980 draws a picture of a lot of progress done but fundamentally the same context, briefly: a recognition of aquaculture as a legitimate sector and therefore user of land and other resources, with leaps in productivity from technological improvement and manpower development, better sharing and application of technology, and better understanding of and sharper tools to tackle problems having complex social and economic-technical-environmental links. But still beset by poor planning and management, diseases and environmental deterioration, public health threats, conflicts among common users, and degraded natural resources. These basic problems impact on the entire society but, as they had always done so, first and foremost on the poor".

O. Genetic in Modest Aquaculture

Drs. P.V.G.K. Reddy and S. Ayyappan's presentation in Fishing Chimes (*Source*: Nov. 1999, page 19) stress the importance of running water fish culture, cage culture, importance of genetics, production and evaluations of several intergeneric and interspecific hybrids etc. which speak on the similar tune the Prof. worked and visualised earlier. An excerpt from their paper is reproduced with some modification.

During the past two and a half decades aquaculture production in India has demonstrated a several fold increase in per ha production (to the tune of 10-15 t/ha/yr.) through the development and adoption of new technologies. These extensive and intensive fish culture practices are essentially based on scientific reasoning and the exploitation of the biological potentialities of the candidate species together with a package of management practices. However, for imparting a further boost to fish production in India it is necessary to explore other avenues. Notwithstanding the experimental work that has been carried out on flow- through (running-water) fish culture and cage culture, further study is needed to develop a standard and sustainable package of technologies to boost fish production under the two systems mentioned above which appear to be promising. Cage-culture, if developed properly, may help in raising the production, particularly from the vast areas of reservoirs of the country.

Another potential means for enhancing fish production during the coming decades appear to lie in the expectation of the genetic potentialities, of our economically important species. As demonstrated in agriculture and in veterinary sector, genetics have a potential role to play in aquaculture too. The ultimate goal being to improve the per capital availability of fish protein to our masses, developmental efforts should aim at improvement and increment of quality as well as quantity of aquaculture products.

Although a beginning has been made in this direction as early as in 1960s by producing and evaluating several interspecific and intergeneric hybrids, serious attempts have been initiated in this regard only from early 1980s. Protocols have been developed to produce highly inbred lines of gynogens (all females) and polyploids (triploids/tetraploids) of Indian major carps as well as Chinese carps through genome manipulations (chromosomal engineering).

These highly homozygous inbred lines (gynogen clones) may have several applied values in modern aquaculture. By top-crossing selected individuals from these lines with those of normal heterozygous ones of the species (intraspecific—cross), heterosis effect can be expected, especially for growth. Gynogenesis can be also used as a potential tool for producing monosex (all females where female homogamety exists). In certain species including Indian carps, where females weigh heavier than males of the same age group, it may be quite possible to enhance the production levels.

On the other hand, triploidy can be utilised in several ways, depending on the species. Triploid/aneuploid individuals usually turn out to be sterile. Sterility in certain species like *Cyprinus carpio* or *Tilapias* which have prolific pond breeding habit can check the overpopulation of the culture medium. Overpopulation in a culture system, particularly in polyspecies culture causes imbalance of the density in the pond. Consequently, the growth of the species in question as well as other species present in the culture system may be effected due to the obvious reason of competition for food as well as space. In certain species of fish, triploidy may also help in enhancing growth, as all the energy is available for somatic growth. Significantly higher growth was reported in triploids of certain species of European and African catfishes and also common carp.

Beside these chromosomal engineering techniques, genome manipulation through gene transfer (genetic engineering), aiming to produce transgenic fish, is another aspect to hold much hope on genetic potential. Recent research efforts elsewhere in the world too have sent signals of a greater promise in this area to produce individuals that grow much faster, resist diseases and withstand environmental stress etc. Atlantic salmon is a good example in this regard. In India also research work in respect of application of genetic engineering in this direction is being pursued.

Selective Breeding

There is also the other approach known as selective breeding which enjoins the hatchery managers (fish seed producers) to follow correct breeding procedures for producing genetically improved quality seed with better survival and growth, and ability to resist disease incidence and withstand environmental stress. Though selection is a longer process to build up better qualities in an individual, its advantage lies in the accumulated nature of the genetic potency from generation to generation. Selective breeding has already demonstrated its potential in several species of fish.

Chapter 5

Prof. Hiralal Chaudhuri's Role in Network of Aquaculture Centre in Asia and the Pacific Region (NACA)

Aquaculture Training

A RESUME

Since his success in induced breeding by hypophysation of pituitary gland in 1957 Prof. Chaudhuri's further research and training of officials have been an inseparable schedule of his duties. During his work in Burma from 1966 to 1970, thereafter in Lao PDR from 1979-84 and SEAFDEC from 1976–79 and again in 1985-88 training of personnel was a part and parcel of his project work schedule.

But most significant part of training of experts from all over the world was as a coordinator in SEAFDEC in various training programmes.

Prof. Chaudhuri acted as NACA Training coordinator. This chapter gives a brief review of his involvement and some recommendations.

To cut short his elaborate activities and involvement in training this editor intends to present an efficient training course drawn by him in 1987 as a part time consultant in an abridged form only:

1. Terms of Reference

The terms of reference of the Consultancy (on a part time basis) were as follows:

☆ To assist the Project Coordinator of NACA in implementing the NACA Training course in the Philippines and to maintain close contact with him.

☆ To organize and implement this Training Course in close consultation with the University of the Philippines in the Visayas and SEAFDEC Aquaculture Department.

☆ To participate in the theoretical and practical instructional work with faculty selection and curricular review.

☆ To assist the Project Coordinator in maintaining accurate financial records of transactions with regards to the payment of training costs and expenses of the trainees.

☆ To prepare a final report on your duties and a fiscal report on accounts.

☆ To support the principles and directions of NACA.

2. Development of the Project

Although the actual consultancy period started from 1 May 1986, Dr. Hiralal Chaudhuri was assigned to take the responsibility of organizing and implementing the 6th Training courses for the Senior Aquaculturists which commenced on 13 March 1986. His appointment letter from FAO was issued on 26 May 1986 initially for 2 months and then for 7-1/2 months after a gap of 2-1/2 months. FAO also assigned him for a full time consultancy for 2-1/2 months (in between Philippines mission) to the Project LAO/82/014 at Vientiane, Lao PDR from 1 July 1986 to advise, coordinate and plan out applied research programmes in the country and in-depth studies on integrated fish/livestock/crop production systems. Dr. Chaudhuri on completion of the assignment prepared a report on "Applied research investigation and technology development in fish culture for Lao PDR" and submitted to FAO and UNDP. The Consultant returned to Tigbauan, Iloilo on 18 September 1986 to continue with the NACA Consultancy.

Prior to his departure to Vientiane, Lao PDR, the Consultant made all arrangements for the overseas study tour of the participants of the 6th Training Course for about 2-1/2 months to Luzon, India, China and Thailand. The trainees left on 2 July 1986 and returned on 10 September 1986 and so the project training activities did not suffer because of the consultant's absence from Tigbauan.

The Sixth Training Course was completed on 6 March 1987 when all the 19 participants were awarded the Diploma in Aquaculture jointly by NACA and the SEAFDEC Aquaculture Department during the Closing ceremonies held on 6 March 1987. Of the 19 participants, 18 qualified to seat for the UPV comprehensive examinations held during 2-4 March 1987 and all of them came out successful and were awarded the degree of Master of Aquaculture by the University of the Philippines in the Visayas.

The consultant completed his assignment on 30 April 1987.

3.　Programme Implementation

Opening Ceremonies and Start of the Sixth Training Course

The Sixth Training Course officially began on 13 March with the Opening ceremonies held at the SEAFDEC Aquaculture Department in Tigbauan, Iloilo. Dr. Ulrich Grieb, FAO Country Representative gave the keynote address. Mr. Pedro G. Padlan was the acting NACA Training Coordinator. Nineteen participants from 8 countries in Asia joined the 6th Training Course. They were initially given an orientation of Tigbauan Research facilities and enrolled at the UPV for the summer semester. A Precourse Examination was given to assess the participants' basic knowledge on various fields of Biology, Chemistry, Fisheries and Aquaculture. On the performance of the precourse examination, each participant was given proper guidance to improve his/her proficiency and knowledge on the subjects including their standard of English. The regular lectures started from 18 March 1986.

Background Information on Course Participants

Relevant background information on the course participants were compiled and distributed to instructors so as to facilitate efficient monitoring of their academic performance and to help them in acquiring the necessary technical know-how.

The Participants' Profile

The participants' profile (below) shows that out of 19 trainees, 2 are from China; 7 are from South Asia (4 Indians, 2 Sri Lankans and 1 Pakistani) and the rest 10 are Southeast Asians (4 Filipono, 3 Thai, 2 Indonesians and 1 Malaysian). There are 8 females which is the highest female proportion as far since the NACA Training Course started. Their academic background shows that 15 are BS and 4 are SM degree holders.

Fellowship and Funding Agencies

The following is a list of the names of the participants, the countries they represent and the Funding Organizations sponsoring their Fellowship:

Sl.No.	Name of Participants	Country	Funding Agency
1.	Baruah, G.C.	India	NACA (FAO/IJNDP)
2.	Chaubal, S. V.	India	-do-
3.	Duangrat Dhesprasith	Thailand	-do-
4.	Fang Xiu Zhen	China	-do-
5.	Dadang Gunarso	Indonesia	-do-
6.	Kathamuthu S/O Subramaniam	Malaysia	-do-
7.	Pholphot Kittisuwan	Thailand	-do-
8.	Rumpon, Danilo P.	Philippines	-do-
9.	Irshad Hussain Shah	Pakistan	-do-
10.	Vinci, G.K.	India	-do-
11.	Hettiarachchi, A.	Sri Lanka	Asian Development Bank
12.	Shantha, K.B.	Sri Lanka–	-do-
13.	Amar, Edgar C.	Philippines	IDRC
14.	Bolivar, Remedios B.	Philippines	-do-
15.	Ladja, Jocelyn M.	Philippines	-do-
16.	Luanprida, Somsak	Thailand	-do-
17.	Shen Ling	China	-do-
18.	Singh, Hardial	India	-do-
19.	Tatie Sri Paryanti	Indonesia	-do-

In addition, one Bangladesh participant Mr. Abdul Khaleque from the 5th batch (1985-86) also attended the summer courses to complete his academic requirements. He was sponsored by COMSEC.

Registration with UPV

Seventeen out of 19 participants were admitted to the Master of Aquaculture programme of UPV for the summer semester. In the case of two, the enrollment was deferred pending submission of original documents. However, all of them qualified for registration for the First and Second semesters.

Selection of Instructors

During early March 1986, the Academic Planning Committee (APC) consisting of 6 members (2 from FAO, 2 from SEAFDEC and 2 from UPV) met at UPV for selecting the Main Instructors of the 17 Courses. Subsequently, also the Committee met for this purpose. The names of the Main Instructors, their affiliations and the corresponding courses were furnished. The UPV Coordinator to NACA took the responsibility of getting the main instructors approved by UPV for their appointment as Faculty members which was a requirement. Other instructors and resource persons were also tentatively selected and invitations were sent to them. Since the training course offered 18 diversified subjects, the host institutions could not provide all the lecturers and hence a number of resource persons were invited from outside. A total of 53 instructors, practical assistants and resource persons took part in the training programme, of which 29 were SEAFDEC AQD staff, 10 from UPV and the rest from FAO and other Organizations and Institutions in and outside the Philippines. Besides, quite a substantial number of instructors and practical assistants from the Regional Lead Centres in India, Thailand and China also contributed their share in the training programme.

Curriculum Development and Lecture Topics

A curriculum for the NACA Training Programme was developed by FAO Task force of experts covering all essential aspects of aquaculture. The task force recommended 337 hours for theoretical instructions and 1,252 hours for practical. This was however, slightly modified and expanded in order to place appropriate emphasis on

Aquaculture practices that have immediate application to the Region. Besides, since the project has an agreement with UPV for the award of a degree in Master of Aquaculture to qualified participants, the course curriculum had to be adjusted to satisfy the academic requirements of UPV and at the same time maintaining the training objectives.

Lecture Topics

During the 6th session (March 1986-March 1987), 17 courses were planned out and implemented. The lectures (in hrs.) cover about 26 per cent and the practical 74 per cent. The consultant's share of lecture hours in the entire course was over 70 hrs.

Laboratory/Practical Topics

During the course of the entire training more emphasis was given to laboratory and fieldwork and almost three-fourths of the total training instruction time was utilized for practicum. Direct exposure and experience in the field gave a better appreciation and understanding as well as conceptualization of related activities as compared to the classroom sessions.

Technical Skill Development

In addition to regular lectures and practicum conducted by the subject-matter specialists and instructors, the participants of the 6th training course were subjected to regular class discussions on specific subjects with the instructors and also held seminars and workshops occasionally on specific topics. In addition, the trainees were asked to prepare term papers on selected topics and gave lectures in the class on term papers prepared by them for developing their technical and communication skills. Also group discussions and self-learning systems were generally encouraged.

4. Seminars and Workshops

The participants were initiated to take active part in regular class discussions and also held regular seminars/workshops on specific topics. A number of seminars were organized under the leadership of the consultant or the instructors of the respective courses where the trainees were assigned topics for oral presentation to develop their ability to stage presence, delivery and their knowledge on the topics. The entire class was encouraged to participate in the discussions.

5.   Field Trips and Study Tours

Field Trips

Iloilo

A number of field trips were undertaken by the participants to several private fish farms in the Iloilo Province to study the fish farming activities. They visited the milkfish/prawn farms (extensive farming) of Ms. Jamandre at Dumangas, of Ms. Jaranilla at Villa and that of Ms. Trespeces also at Villa. Besides, they visited the finfish floating cages at the Igang Sub-station of SEAFDEC at Guimaras, Iloilo.

Aklan and Capiz

The participants visited the private shrimp hatcheries, oyster and mussel farms at Batan, Aklan and Roxas City, Capiz. They also observed other aquaculture activities in those areas during their two days trip on 4-5 October 1986.

Cebu and Bohol

The participants undertook a four day trip (21-25 October, 1986) to Cebu and Bohol Provinces in the Central Visayas to plant marine seaweeds in seaweed farms and also visited the seaweed processing plants in Cebu.

Study Tour

The participants undertook study tours both in the Philippines as well as to several Asian countries. The study tours provided opportunities to study various aquafarming systems as practiced in countries visited, so as to widen their vision and knowledge of aquaculture.

Luzon Island (Philippines)

A seven-day study tour (2 July–8 July 1986) was undertaken to various important aquaculture centres (mainly freshwater) in Luzon Island in the Philippines before the participants' departure for the overseas study tour. The aquaculture centres visited were Binangonan Research Station of SEAFDEC Aquaculture Department, Central Luzon State University (CLSU), Laguna Lake pen and cage culture for milkfish and tilapia and tilapia cage culture in Lakes Bunot, Sampaloc and Calibato. They also visited fish farm at Tanay,

Rizal and also the office and library of ICLARM at Makati. Dr. Crispino Saclauso, UPV Coordinator and Ms. Morena Fernandez, Trainee Affairs Assistant, accompanied the trainees during the tour.

Overseas Study Tour to India, Thailand and China

As in previous years, an overseas study tour was organized to the other three regional centres in India, Thailand and China for 17, 21 and 30 days respectively. These Asian countries have long experience in aquaculture. The overseas study tour (8 July–10 Sept. 1986) mainly included activities such as, study of various fish culture and pond management systems, seed production and hatchery management, integrated fish/livestock/crop farming systems, composite fish culture and polyculture practices, primary productivity studies, aquatic micro-biology, aquaculture planning and extension, production economics and aquaculture industry development policy and administration in the countries visited. The main emphasis, however, was given to the following studies:

☆ Large-scale seed production by induced breeding and other methods of commercially important species such as, Indian and Chinese Carps, Catfish (*Clarias* and *Pangasius*), giant freshwater prawn (*Macrobrachium*) and others.

☆ Integrated fish/ livestock/crop farming systems and their economics.

☆ Different aquafarming systems in the countries visited and their comparative merits.

☆ Aquaculture research and development: its organization and administration.

☆ Study or reservoir fisheries and their management.

During the tours the participants visited many important Research Institutes and Research Centres in the Countries visited.

6. Completion of Courses and Exams

All the 17 courses were completed by the second week of February 1987 and the participants were given more than two weeks time for recaptulation for their Comprehensive Exam. The academic performance of trainees was assessed by continuous assessment through periodic examinations, field and laboratory reports, term papers and practical work. The final examination for each course including practicals were held as soon as the courses were completed. The grades are given according to the grading systems of the University of the Philippines in the Visayas. A weighted average grade of 2.0 is the minimum for the eligibility to seat in the Comprehensive Exam for the degree in Master of Aquaculture by UPV. The grades (weighted average) for all the 19 participants were furnished. The grade list shows that only Mr. Dadang Gunarso from Indonesia (INS/81/008) got a grade slightly lower than 2.0. In a meeting with a panel of instructors held on 24 February 1987 for recommending participants for the diploma in Aquaculture to be awarded by NACA and SEAFDEC, all the 19 participants were recommended for the diploma. The case of Mr. Dadang Gunarso was especially taken up. Since in previous years a grade below his was considered for the diploma, the panel recommended him eligible for the diploma.

The examination results of the participants showed that the 6th batch obtained a weighted average of 1.63 as against the 5th batch of.1.72. One participant of the 6th batch obtained an all time highest weighted grade average of 1.1875 and three others around 1.34 to 1.36.

Comprehensive Examination and Award of M.Aq. Degree by UPV

The comprehensive examination for the 6th batch NACA Training Course was held during 2-4 March 1987. An Examination Committee was organized with Prof. Valeriano Corre (UPV) as the Chairman. The other members were Dr. Crispino A. Saclauso (UPV), Prof. Virgilio A. Dureza (UPV), Ms. Jurgenne H. Primavera (SEAFDEC) and Dr. Hiralal Chaudhuri (NACA/SEAFDEC). The exam was divided into three parts: the general field (45 per cent) was held on 2 March 1987 and the specialized field (45 per cent) on 3 March and the oral exam (10 per cent) was held on 4 March 1987. Excepting Mr. Dadang Gunarso, all the 18 participants were qualified to seat for the Comprehensive Examination. However, at the request of the NACA Training Coordinator, Mr. Dadang Gunarso was permitted to seat for the said exam for NACA purposes only.

The results of the comprehensive examination was very encouraging and all the 18 qualified participants came out successful. Dadang, however, could not get the passing mark. The highest mark obtained by a participant was 86.61 per cent whereas the average for the batch was 80.94 per cent as compared to last years' (5th batch) average of 75.83 per cent.

The University of the Philippines in the Visayas has awarded the degree in Master of Aquaculture to all the 18 NACA 6th batch participants in the Graduation Ceremony held on 2 April 1987.

Closing Ceremonies and Departure of the Participants

The closing ceremonies were held on 6 March 1987 at the Amigo Terrace Hotel, Iloilo City at 5:30 p.m. Dr. Ulrich Grieb, FAO Representative, Manila gave the keynote address and Dr. Flor J. Lacanilao awarded the diplomas. Mr. Turhan Mangun, UNDP Resident Representative, Manila accepted the invitation to give a speech as the guest speaker but at the last moment he could not come but his prepared speech was read by Mr. N. Brown, Asst. Resident Representative, UNDP. Dr. Dionisia Rola, Chancellor of UPV also delivered a speech to the participants as a guest speaker. Since Director Juanito Malig of BF AR could not attend, his speech was read by his representative, Mr. Justo Montemayor, Asst. Director of Fisheries. All the 19 participants received the Diploma in Aquaculture. The ceremony ended with a dinner.

7.   Additional Support to Project Activities

Short-term Training Programme

A training course on small-scale Prawn-hatchery Operations and Management was organized by the SEAFDEC Aquaculture Department with NACA support. Thirteen NACA-funded trainees participated in the training course out of a total of 39 trainees. The course was held for 7 weeks during 4 August to 25 September 1986.

Philippine National Aquaculture Centre (TCDC Programme)

BFAR-FAO/NACA Shrimp Hatchery Project (TCP/PHI/6651) for strengthening of the Philippine National Aquaculture Centre (PNAC), stationed at Calape, Bohol is linked with RLCP. According to the agreement, SEAFDEC Aquaculture Department as the Regional

Lead Centre in the Philippines will provide technical assistance and training of the PNAC technicians. During 7-9 April, 1987 Mesrs. Antonia Villaluz and Elmer Sumbing of SEAFDEC visited the proposed centre and gave their recommendations (dated 25 April 1986) based on their observations and attached a list of the life support systems and other equipments needed to complete the hatchery and make it operationalized. Subsequently, Mr. Villauz visited two more times and gave his advice and recommendations (dated 13 Nov. 1986 and 16 March 1987) for the modifications and completion of the Shrimp Hatchery.

At the request of the NACA Project Coordinator, the Consultant visited Cebu-Calape, Bohol during 6-9 April 1987. He observed that the 2.9 ha grow out pond area has been prepared and are now ready for stocking with shrimp fry. As regards the shrimp hatchery, the modifications, construction and installation of the seawater component as recommended by Mr. Villaluz, have been completed. As for the aeration component, the roots blower could not be procured as yet. The purchase order was being made through FAO, Rome. However, some details of the specifications of the roots blower were missing. The consultant on his return obtained the specifications from Mr. Villaluz and communicated the same to Dr. Grieb, FAO Representative for onward transmission to Rome. The hatchery would be operational as soon as the roots blower was procured and installed.

8. Brief Review of the Training Programme and Recommendaions

Brief Review of the Training Programme

From a review of last 6 years' Training Programmes it could be definitely concluded that the programme was able to achieve considerable success. The broad-based and multi-disciplinary approach of this Training Programme has immensely benefited the participants and contributed significantly to strengthen and widen their knowledge in aquaculture. There has been a regular feedback also from the participants of the last 6 Training Courses as they were asked to evaluate and assess each course and organization of the courses including the assessment of the instructors. In general, the courses have received great appreciation and favourable comments from the participants as reflected in their assessments.

The participants did learn a lot about all aspects of aquaculture. In addition, the overseas study-tours, seminars, class discussions, development of communication skills and leadership training have helped the participants to be self-confident in expressing their ideas and developing a sense of leadership in them and also in establishing linkages with different people with different socio-economic and cultural background from the region. However, there had been complaints of the courses being "emotionally, intelectually and physically demanding", having ultra-heavy loads thereby producing lot of tensions and sleepless nights especially before the exams. One of the batches after complaining of tension remarked, "All our activities and even our lives were centered on this training course". However, at the end of their mentally exhaustive experience with all the tensions, cramming and sweat, most of them were happy to take home with them an achievement, a Master's Degree in Aquaculture.

The Consultant was aware of the above facts and he tried his best to lessen the tension by giving the participants intermittent free time as far as practicable. That the course curriculum was too tight and workloads heavy, was no doubt understandable because two years' time meant for the degree in M. Aq. had been clamped to one year's course. It is quite possible that because of the mental stress and too much heavy pressure of work for the M. Aq. programme, the participants work hard to pass and not necessarily learn as is expected.

The Consultant reviewed the pros and cons of having a degree or a non-degree course, observed that the degree course has several other disadvantages.

NACA Training Course time schedule does not correspond to the regular time schedule of UPV and for the past 6 years UPV had to allow enrollment for the NACA candidates in March which is not in conformity with the University policies. UPV therefore, would like NACA Training Programme to start coinciding the regular school semesters, either in June or in November. These timings again do not suit NACA since the overseas study-tour which is an essential and indispensable component of the course and scheduled from July to September to coincide with the breeding seasons of the cultivated fishes in the countries visited, cannot be changed. Besides, the trainees need some time to be taught the basic principles of aquaculture and a few other introductory courses before their departure for the overseas tours, in order to fully appreciate the various aquaculture practices in other lead centres. This excludes the time needed to process their official papers and travel documents necessary prior to departure.

The participants in their assessment considered many courses too elaborate and time consuming. The Consultant also feels that at least the following three courses (*a*) Fish Nutrition and Artificial Feeding, (*b*) Post-harvest Technology and (*c*) Fish Health and Sanitation are too heavy and elaborate. Senior Aquaculturists who are not interested to specialize in these subjects need not go into so much details. In fact, these courses in UPV are full-time semester courses and are completed at NACA in 3 weeks only. Hence, in every year the performance of the participants' are not very satisfactory in these 3 courses.

Further, many of the courses do not cater to the needs of all the countries in Asia especially, the land-locked ones such as, Nepal, Bhutan, Laos and Afghanistan who are not interested in courses involving marine and brackish water culture practices. On the contrary, in a non-degree training course emphasis could be given to courses beneficial to all the countries in the region and giving them option to specialize in any subject in which the particular country is interested. However, in case some Governments want degrees, the participants could be admitted to the regular degree course of UPV being sponsored by NACA.

The present training course is mostly oriented towards brackish water and mari-culture aspects of aquaculture. It could be observed from the participants' profile of the 6th NACA training batch that over 84 per cent of the participants were from freshwater discipline and majority of them are expected to continue their work in freshwater areas.

Another important consideration against the degree course is the poor knowledge in English of participants from several countries. This is a big handicap for these trainees and for the instructors as well. Since the entire training programme including the courses is in English language the trainees feel miserable at the beginning. Although they try to learn by engaging tutors but because of the tight schedule of the training programme they hardly find sufficient time. This problem should be taken into account in case degree programme is continued.

Lastly, there is another problem that some Governments may have to face is that trainees after obtaining their M. Aq. degree may demand for promotions.

9. Recommendations

The Consultant after reviewing the Training Programme and from his personal experience as the NACA Training Coordinator implementing the 6th NACA Training Course for Senior Aquaculturists in Asia and the Pacific, Region offers the following recommendations:

The Training Programme may continue at RLCP in SEAFDEC Aquaculture Department at Tigbauan, Iloilo preferably as a non-degree one-year training course with thorough reorganization of the existing training courses and programme. However, those Governments who are in favour of a degree may get their nominees admitted to the regular Master of Aquaculture programme of the University of the Philippines in the Visayas and NACA (FAO/UNDP) Project may fund the nominees and their stipends.

The non-degree training course curriculum should have three component parts: (*a*) A period of 6 months or so for lectures and practicals for all the general courses with emphasis on all important aspects of aquaculture (*b*) A period of 2-1/2 months for overseas study tours and (*c*) the remaining 3-1/2 months or so for specialization on any field of interest.

The overseas study tour is essential and should continue with itinerary suitably revised. The number of days tour in India should be increased. Instead of visiting Central Institute of Freshwater Aquacutlure (CIFA) at Dhauli for most of the time, tours should be conducted also to a few other important aquaculture centres (fresh, brackish and marine) in India and also to one or two reservoir fisheries and visits to the Central Institute of Fisheries Education (CIFE)/College of Fisheries.

The Specialized (elective) courses may include the following: (*a*) Fish Nutrition, (*b*) Prawn Hatchery Operations and Management, (*c*) Finfish Hatchery Operations and Management, (*d*) Freshwater Fish Culture, (*e*) Integrated fish/livestock/crop Farming Systems, (*f*) Post-harvest Technology, (*g*) Fish Health and Sanitation, (*h*) Pen and Cage Culture, (*i*) Aquaculture Economics and others.

Some of the existing courses should be reduced or modified suiting the time allotted for each course. Some courses on the other hand should be strengthened such as, Aquaculture Planning and Aquaculture Extension with management subjects to be added. And lastly, new courses like Statistics and Computer Programming may be introduced.

In case the present M. Aq. degree training course is retained, then it may be necessary to have adjustments of the programme to conform with the University Academic Calendar activities.

With regard to admission to the Masteral degree programme of UPV, the notice regarding all essential requirements for enrollment should be forwarded to the participants well in advance and to avoid complications in registration all these requirements should be submitted to UPV positively before the commencement of the training course.

The applications for the nominees (degree or non-degree) should be scrutinized by the nominating Governments to ensure that only suitably qualified personnel are selected for the training programme.

All nominees to the training programme (degree or non-degree) should be proficient in English language.

Otherwise, they should take up English classes for at least 6 months before they join the training courses. This is essential for improving the quality and standard of the entire training programme.

Chapter 6

Break-through in *Chanos chanos* Induced Breeding: Another Feather in Prof. Chaudhuri's Cap

A RESUME

The readers may refer to the editor's expositions in chapter–I on the predictions of famous aquaculturits about the golden finger of Prof. Chaudhuri in the laboratory and field research which could solve the issue of induced breeding of milkfish.

Only a few of his papers were available and mentioned hereinafter to register that the prediction made in 1957 and thereafter, became a reality after 20 years when the professor got the first opportunity to work on milkfish.

It may be realized that this species is choicest of the principal fish food of people of Taiwan, Philippines, Indonesia and very may maritime pacific ocean countries.

In fact his work has been a revolution in biology and fish culture.

The papers discussed are:-

A. Induced Breeding Chanos chanos

B. Use of hormones in breeding cultivated warm-water fishes with special reference to milkfish.

Fruitful predictions made in late fifties and sometime later–about milkfish induced breeding.

1. Views of Experts

Science 1957 in after the successful induced breeding of major carps some world famous aquaculturists made the following predictions.

- ☆ Dr. S. Sarig, Aquaculturist of Israel wrote to Prof. Chaudhuri in 1958 that "If you succeed in breeding milkfish *Chanos chanos*, it would be a revolution in Biology and Fish Culture".

- ☆ Dr. D. D. Moss, famous Aquaculturist of USA–wrote to him after the professor had joined SEAFDEC (1976) that–

"If any body world accomplish this task of induced breeding of milkfish Prof. Chaudhuri would be the one"

Dr. Moss further wrote (1977) that–"We have been keeping up with your progress in breeding of milkfish". In fact Prof. Chaudhuri did not get the opportunity to do research on milkfish as the International Development Research Centre (IDRC) of Canada is initiated a project under WE van store and other aquaculturists.

. But in Tigbauan Prof. Chaudhuri continued work on this fish and on his initial research success Dr. Moss wrote–

"It is very encouraging that the eggs of *Chanos chanos* which were collected undoubtedly under confinement have been grown to post larval stage. This fact itself is a world record in the strict sense and a significant step towards the goal. I am glad to find this record established in the Philippine before Hawaii and Taiwan. Sexual dimorphisms of *Chanos* through your hand and Biology was made clear".

This letter was written after Dr. Chaudhuri attained success in artificially fertilizing milkfish (*Chanos*) eggs caught from the sea and studying the development and rearing upto the post larval stage. He published the paper in the journal. 'Aquaculture' published from the Netherlands (1977). He had also succeeded in determining the sexes of milkfish which were unknown till then.

2.   *Chanos chanos* Forskal (Milk Fish)

The milkfish *Chanos chanos* (Forskal) is the most extensively cultivated in coastal ponds in the Pacific Ocean maritime countries. For centuries larvae of this fish have been collected from coastal areas and reared in brackish water fish ponds and fish pens. It is an important food fish all over the region.

Yet very little has been known about the stages of its developmental history in the wild. It does not attain maturity or breed in captivity. The eggs hatch within 12 hours of collection but the larvae die. So urgent need was felt to develop a dependable source of seed. Fishery scientists in a number of countries were engaged in that research on inducing maturation and spawning in both pond-grown and wild milk fish by hypophysation.

Excerpts of the Professor's publication are presented in this treatise. The title of the publications are:

☆ "Notes on the external sex charachers of *Chanos chanos* (Forskal) spawners" by Chaudhuri, Juario, Samson and Tiro (1976).

☆ "Use of hormones in breeding cultivated warm-water fishes with special refocuses to milkfish, *Chanos chanos* (Forskal)", by Chaudhuri and Juario, (1977).

☆ "Artificial fertilization of eggs as early development of the milkfish *Chanos chanos* (Forskal)" by Chaudhuri (1977)

3.   External Sex Characters of Milkfish

The professor made the following observations which are put forward mostly in his own language.

ABSTRACT

In the present study, no visible differences between the sexes of the *Chanos chanos* with reference to external features such as coloration, shape of head, snout and operculum, presence of tubercles or nasal pores, length, size and shape as well as any roughness in the various fins, could be found. However, the anal region of the mature milkfish (sabalo) exhibits discernible anatomical differences in the male and female.

The male has two main openings visible externally: the anterior anus and the posterior urogenital opening at the tip of the urogenital papilla.

The female has three main openings instead of two: the anterior most anus, followed by the genital pore and the urinary pore located posterior to the genital pore at the tip of the urogenital papilla.

Internal examinations were also made on both sexes.

In ripe sabalo, it is easier to distinguish the sexes since milt oozes out of the urogenital pore by pressing the abdomen of the ripe male fish. Gravid females are identified by their distended abdomen.

About the Fish

The milkfish, *Chanos chanos* (Forskal), has a wide distribution and is considered to be an important food fish in Southeast Asia, especially in the Philippines, Indonesia and Taiwan where the fish is cultured both the brackish as well as in freshwater environments. Since the fish does not mature in ponds, the only source of procurement of its seed is the sea. The collection of milkfish fry (10-15 mm) from the coastal waters is an age-old practice in countries where it is cultured. However, with the expansion of milkfish culture the demand for its seed has increased considerably and the seed thus captured from the sea is too inadequate to meet the increasing demands of the industry. Hence, concerted efforts are being made in several countries to induce the captured milkfish from the wild to breed by hormone injections.

The first breakthrough in spawning milkfish was recently achieved at the Aquaculture Department of SEAFDEC when eggs obtained from injected females were successfully fertilized by sperms from injected males (SEAFDEC Progress Report, 1977). The eggs hatched out successfully and several larvae were obtained. The technique, however, has to be standardized so as to produce seed commercially, and successfully rear them for stocking in ponds.

One of the main constraints in the studies in breeding milkfish is distinguishing the sexes externally. The milkfish spanners are large, very active and need very careful handling to avoid stress or injury to the animal. It is, therefore, essential that a practical method of differentiating the sexes externally be found to facilitate spawning of the species for large-scale production of seed. The present note incorporates the observations made on the external determination of sex in mature milkfish or Sabalo based on the characters of the anal region of the fish.

Observations

"In our present study, we did not find any visible differences between the sexes with reference to external features such as coloration, shape of head, snout and operculum, presence of tubercles or nasal pores; length, size and shape as well as any roughness in the various fins. However, the anal region of the mature milkfish (Sabalo) exhibits discernible anatomical differences in the male and female."

– Says Professor Chaudhuri

In the males, there are two main openings visible externally. These are the anterior anus and the posterior urogenital opening at the tip of the urogenital papilla. Internally, the vasa deferentia (male genital ducts) from the testes join into a common duct about 5-10 mm from the urogenital pore. The urinary pore opens into this common duct from the dorsal side. In addition, there are two small pores situated on each side of the base of the urogenital papilla opening ventrally into the coelom.

In the females, there are three main openings in the anal region instead of two as found in the males. The anterior-most opening is the anus followed by the genital pore. The third opening is the urinary pore which is posterior to the genital pore located at the tip of the urogenital papilla. On internal examination, it was observed that the two oviducts join to form one common broad oviduct which finally opens through the genital pore. In females just as in males, there are two lateral pores at the base of the urogenital papilla opening into the body cavity. The function of these abdominal pores is not known. Similar abdominal pores are also present in sharks and skates. According to Hyman (1942), in spiny dogfish "the pleuroperitoneal cavity communicates with the exterior by means of the abdominal pores. These will be found one on each side of the anal opening (in the skate posterior to the anus), somewhat concealed by a fold or skin. Their purpose is obscure".

In ripe Sabalo, it is easier to distinguish the sexes since milt oozes out of the urogenital pore by pressing the abdomen of the ripe male fish. The gravid females are identified by their distended abdomens.

Male Sabalo are usually smaller in size than females. The weight range of specimens examined at the Tigbauan Station of the SEAFDEC Aquaculture Department varied from 3.5 to 7. 0 kg in males and 4.5 to 10. 7 kg in females. Specimens studied at the Station at Pandan, Antique in 1975 showed that the males varied in weight from 4.1 to 7.4 kg whereas the females had a weight range of 5.1 to 7.7 kg (Tiro, et. al., 1976).

It was observed that the male Sabalo moved faster while the ripe females were comparatively sluggish in movement. The maximum weight of 1.3 kg of ovaries was measured in one female fish weighing 10.7 kg, the GSI being 12.1 which appears to be the highest value of GSI so far reported in milkfish. In another specimen having a total weight of 8 kg, the ovaries weighed 965.3 g and the GSI was 12.07.

This important study on the differentiation of sexes of milkfish based on the external characters is being continued. Our present observation on the anal region of adult milkfish (Sabalo) has been used to sex over 60 specimens with 100 per cent accuracy. In fact, this method of sexing has been used regularly in our milkfish breeding experiments leading to successful fertilization of milkfish eggs at Pandan and Tigbauan. It is felt that the method is practical and accurate for field purposes in sexing milkfish spawners.

References

Chaudhuri, H. (1959): Notes on the external characters distinguishing sex of breeders of the common Indian Carps. *Sci. and Cult.* 25:258-259.

(1962): Induced breeding and development of common catfish, *Ompok bimaculatus* (Bloch.) *Proc. Ind. Scl. Cogr.* 1962.

Hyman, L. H. (1942): *Comparative Vertebrate Anatomy*, University of Chicago Press, Chicago 37, 1942, pp. 544.

Mane, A. M. (1929): A preliminary study of the life history and habits of kanduli (*Arius* sp). in Laguna de Bay. The Philippine *Agriculturist*, 18(2):81- 118.

Norton, V. J., H. Nishimura and K. B. Davis (1976): A technique for sexing channel catfish. *Trans. Am. Fish. Soc.* 3:460-462.

Ramaswamy, L. S. and A. D. Hasler (1955): Hormones and secondary sex characters in the minnow, Hyborhynchus. *Physiological Zoology,* 28(1):62-68.

SEAFDEC Progress Report No.1, May 1977: Background information on Milkfish Research Breakthrough at the SEAFDEC Aquaculture Department.

Tang, Y. A. (1954): On the processes and ridges on the pectoral fin rays of the males of *Hypophthalmithys molitrix* (C. and V.) and *Aristichthys nobilis* (Richard). *Fish. Cult. Rep.* Taiwan, 1:12 p.

Tiro, L. B. Jr., A. C. Villaluz and W. E. Vanstone (1976): Morphological measurements, gonadal development and estimated ages of adult milkfish *Chanos chanos* captured in Pandan Bay from 10 May–16 June 1975. *Proc. Int. Milkfish Workshop Conf.* May 19-22, 1976, Tigbauan, Iloilo, Philippines. Working Paper No. 16:193-203.

Use of Hormones in Breeding Cultivation Warm-Water Fishes with Special Reference to Milkfish, *Chanos chanos* (Forskal)

ABSTRACT

The role of hormones in the controlled reproduction of a few test fishes is well documented. However, information on the mechanisms of endocrine regulation of ovulation in cultivated warm-water fishes is very meager.

Hormones, especially the gonadotropic hormones of piscine origin, are increasingly being used in modern aquafarming to produce the seed of many important cultivated fishes. While chorionic gonadotropin and other exogenous mammalian hormones are used in spawning the channel catfish, fish pituitary hormones are usually needed to induce spawning in the difficult-to-spawn Asiatic carps. In mullets, however, either homo- plastic pituitary gland or human chorionic gonadotropin (HCG) or a mixture of HCG and a threshold dose of the former is ordinarily injected to precipitate spawning. Of late, semi-purified salmon gonadotropin (SG-GIOO) has been used to induce spawning in several species of food fishes.

While several marine fishes have been artificially bred by administration of hormones, induced spawning of the milkfish, *Chanos chanos* has been tried with little success. The milkfish is a widely distributed food fish extensively cultivated in ponds in Southeast Asia. Recently, significant results have been obtained in spawning mature milkfish captured from the wild by hormone injections.

The experiments conducted on induced breeding of milkfish leading to the successful fertilization and hatching of milkfish eggs are briefly described.

1. Professor Chaudhuri's Observation

Adequate supply of quality seed of cultivated species is an essential prerequisite to successful aquaculture. Controlled reproduction is a major problem with many important warm-water cultivated species. The species do not normally reproduce in captivity or do not attain sexual maturation when cultivated in ponds. Hormones are increasingly being used in modern aquafarming in inducing spawning of such species to obtain their seed for culture.

For developing effective methods of seed production at any season, it is essential that the role of the endocrine systems and the endocrine control of reproduction be well understood. While considerable work has been done in recent years on the reproductive endocrinology of several test fishes, information available on the endocrine mechanism of maturation of oocytes and ovulation in important warm-water cultivated species is fragmental.

This paper is a brief review of the use of various exogenous hormones and other effective triggering agents in inducing spawning and obtaining seeds of important cultivated species, namely the difficult-to-spawn Asiatic carps, catfishes, mullets and milkfish. The experiments leading to the first breeding' success with milkfish are also discussed.

2. Use of Hormones in Spawning Catfish

Catfishes are an important group of warm-water cultivated species. Majority of these species do not create any major problem in the production of seed and can be bred by manipulation of the pond environment and by provision of proper facilities for spawning. For certain species, however, hormone is being used either to regulate their spawning or to provide additional stimulus in the form of exogenous hormone. The important warm-water catfish species are *Ictalurus punctatus, Pangasius sutchi, P. pangasius, Clarias macrocephalus, C. batrachus* and *Heteropneustes fossilis.*

In aquarium spawning of the channel catfish Ictalurus punctatus, either fish pituitary extracts or HCG is injected. Females are usually injected with 2.2 to 22 mg of acetone-dried pituitaries administered in three doses or with a single dose of 600 to 2200 IU of HCG/kg body weight. Spawning normally occurs within 16 to 24 hours after the last injection (Bardach, Ryther and McLarney, 1972). *Pangasius sutchi* and *P. pangasius* are regularly bred in Thailand by injection of fish pituitary hormones. In Thailand while *Clarias macrocephalus* is induced to breed by injecting 13-39 mg of homoplastic pituitary glands per fish weighing 90-190 g (Tongsanga, Sidthimunka and Menasveta, 1963), pond breeding of *C. batrachus* is done without hormone injection. *C. batrachus* and *H. fossilis,* the two air-breathing cultivated catfishes have largely been used as laboratory test animals for the study of reproductive endocrinology of fish and considerable information is available on these fishes (Ramaswamy, 1962; Sundararaj and Goswami; 1966).

3. Use of Hormones in Spawning Mullets

Although mullet culture is an age-old practice in many countries, till recently, collection of the fry from the natural habitats has been the only source of seed for culture. Successful induction of spawning of the grey mullet *Mugil cephalous* has been accomplished since the early sixties. While injection of homoplastic pituitary gland extracts (8 to 16 mg/kg body weight) resulted in ovulation of the grey mullet in India (Chaudhuri, 1968), the fish received a dose of 2 to 3.5 mullet hypophysis in combination with 30-40 RU of Synahorin for successful spawning in Taiwan (Tang, 1964; Liao,1975). In Israel, Yashouv (1969) induced spawning of freshwater pond-reared *M. cephalus* by using carp pituitary glands. Induced spawning in Hawaii was attained by injecting 5 to 8 salmon pituitary glands in combination with 175-280 RU of Synahorin (Shehadeh and Ellis, 1970). HCG alone also gave successful spawning when a prime dose of 20 IU/R body weight was followed by a larger dose of 40 IU/ g body-weight at an interval of 26 hours (Kuo, Shehadeh and Nash, 1973). In the Oceanic Institute, Hawaii a practical *in vivo* method was developed to assess the stage of development of Intra-ovarian oocytes. In female *M. cephalus*, HCG (50 IU/500 9 body-weight), FSH + LH (15 ug + 5 ug/500 g body-weight), and mullet pituitary extract † Synahorin (100 ug † 30 RU 1500/g body weight) were used to induce oocyte maturation and prevent onset of atresia (Shehadeh, Madden and Dohi, 1973). Also SG-G100 at 100 ug/kg body weight was effective in inducing ovulation in the grey mullet with mean oocyte diameter of 0.75 mm and natural spawning was induced in all females with a total dose of 11.9 to 2.9 ug/g body weight (Shehadeh, Kuo and Milisen, 1973).

4.   Use of Hormones in Spawning Milkfish

The milkfish *Chanos chanos* is extensively cultured in several countries and is an extremely popular food fish in the Philippines, Indonesia and Taiwan. Since the fish does not develop gonads in captivity, the only source of procurement of its seed is the coastal waters from where milkfish fry are collected for stocking ponds. In order to assure a dependable supply, an urgent need is realized to induce maturation in pond-reared milkfish and induce spawning in wild milkfish spawners.

Liao and Chang (1976) raised adult milkfish for 5 to 6 years under artificial conditions. While the males developed mature testes, the females failed to mature. Some of the ovarian oocytes (0.54 per cent in the 5- year female and 8.41 per cent in the 6-year female) were observed to be only at the oil droplet (yolk vesicle) stage. The experiment was continued with intermittent injection of hormones but no positive result has been reported so far.

Preliminary experiments carried out in the Philippines in the early seventies to induce adult milkfish captured from the sea to spawn by hormone injections were not successful. Single intraperitoneal injection of 10,000 IU of HCG (about 1500 IU/kg. body weight) failed to induce spawning (Anon, 1974). Nash and Kuo (1976) tried SG-G100 in captured milkfish spawners and the same dose (10 ug/g body-weight) as developed for mullet was used. Only one fish which received two injections of 25 mg each of SG-G100 had hydrated eggs and partially ovulated. It appeared that the dose used for mullet was too high for the milkfish and a smaller dose between 12 to 15 ug/g body weight was suggested. It was also observed that eggs having a diameter of 0.8 mm and above became atretic due to excessive handing and stress if no exogenous gonadotropin was introduced within 26 hours.

Initial investigations carried out at the SEAFDEC Aquaculture Department on the artificial propagation of milkfish resulted in the release of hydrated eggs after injection with SG-G100. Milkfish spawners weighing 7 to 8 kg were injected with 30 mg SG-G100 in 3 ml of 1.0 per cent NaCI. One injected female released hydrated eggs (1.18 mm diameter) 17 hours after the first injection. A second female receiving a second injection of similar dose also released hydrated eggs (1.11 mm diameter). In another experiment hydrated eggs were released by a female injected with 50 mg of SG-G100. The hydrated eggs obtained in all cases, however, were unfertilized (Vanstone, *et al.*, 1976).

The first breakthrough in induced spawning of milkfish was recently achieved at the Aquaculture Department of SEAFDEC (SEAFDEC Progress Report, 1977) and successful hatching of artificially fertilized eggs was accomplished on several occasions both at the Pandan and Tigbauan milkfish breeding stations (Chaudhuri, *et al.*, 1977a; Vanstone *et al.*, unpublished). Adult milkfish captured from the sea were used for the breeding experiments. The spawners were sexed (Chaudhuri, *et al.*, 1977b) and held in canvas tanks before and after the hormone injections.

At Pandan a gravid female received 4 ml of hormone solution containing 15 mg acetone-dried Pacific salmon pituitary powder and 100 IU chorionic gonadotropin per ml. A sexually mature male was also injected similarly. A second injection consisting of 6 ml of the hormone solution was given after an interval of 9-hours. Both the injected male and the female were killed and stripped 15 hours after the second injection and the eggs successfully fertilized. A second gravid female was given a dose of 60 mg of acetone-dried Pacific salmon pituitary powder combined with 400 IU of chorionic gonadotropin. Subsequently, two more injections were given with an interval of about 9 hours between injections, and a dose 1-1/2 times more than the previous injection. The injected female was stripped approximately 9 hours after the third injection and the eggs were successfully fertilized (SEAFDEC Progress Report 1977; Vanstone *et al.*, unpublished).

At Tigbauan, the spawners were cathetered with a narrow plastic cannula and a small sample of sex products was collected for examination. The female milkfish is injected only when the intra ovarian eggs were found to be about 0.7 mm in size. The induced breeding experiments conducted on milkfish have indicated that a low dose of chorionic gonadotropin alone up to 1,500 IU (Antuitrin S) has no effect on the gonad of milkfish. Also injection given to females with 725 IU (CG-Ayerst + 19 mg tuna pituitary gland/kg body weight did not give successful spawning in milkfish. A gravid female injected with acetone-dried salmon pituitary gland with a dose of 18 mg/kg body weight, in two injections responded and showed hydration of eggs but the female died due to stress about 14 hours after the second injection (Chaudhuri, *et al.*, unpublished).

5. Conclusion

The experimental results as described above demonstrate that it is possible to induce milkfish to breed by injections of pituitary hormones either alone or in combination with chorionic gonadotropins if gravid spawners are available. Induction of gonadal maturation in milkfish is still a problem. Studies on the reproductive physiology of milkfish have to be conducted especially on the endocrine mechanism of oocyte maturation.

A brief review on the status of the use of hormones in modern aquaculture practices in obtaining quality seed of warm-water cultivated species has been presented in the foregoing. While considerable work has been done on a few species like *H. fossilis* and *C. batrachus* and also to some extent on the grey mullet *M. cephalus*, very little information is available on the hormonal regulation, the mechanism of oocyte maturation, ovulation and spawning of the majority of the important cultivated species. It is essential that the reproductive physiology of these fishes be well understood for inducing maturation of gonads under artificial condition and induction of spawning in these problem species.

While piscine gonadotropin is largely used for spawning in regular aquaculture practices, active research is required to evolve a suitable and more efficient substitute for it. Recent studies on several steroid hormones, synthetic luteinizing hormone-releasing hormone (LH-RH), and several extra-biological substances such as clomiphene citrate, copper acetate and asphalt, clomid, sexovid, etc., have given encouraging results in accelerating oogenesis and vitellegonesis and inducing ovulation in certain fishes. The presence of prostaglandins in teleosts and their possible roles in triggering gonadotropin secretion in pituitary glands are being studied in several laboratories. The use of methallibure, a gonado-tropic inhibitor has thrown a new light on fish endocrinology (Billard, *et al.,* 1970). These studies have opened up new possibilities in furthering our knowledge on the endocrine control of reproduction in fish and could provide better technology towards inducing maturation and ovulation in aquacultured fish.

[Source: H. Chaudhuri and J.V.Juario in
Fishery Research Journal of the Philippines, 1977]

References

Chaudhuri, H. 1968. Breeding and selection of cultivated warm-water fishes in Asia and the Far East: A review. *FAO Fish. Rep.* (44), 4, 30-66.

————1976. Use of hormone, in induced spawning of carp. *J. Fish. Res. Board Con.* 33, 940-947.

————Alikunhi, K.H.. 1957. Observations on the spawning of Indian carps by hormone injection. *Curr. Sci..* 26, 381-382.

————Juario, J.V. Primavera, J.H.. Samson, R., Mateo, R., 197a. Observations on artificial fertilization of eggs and the embryonic and larval development of milkfish *Chanos chanos* (Forskal). *Aquaculture* (In press).

Liao, I.C.. 1975. Experiments on induced breeding of the grey mullet in Taiwan from 1963 to 1973. *Aquaculture,* 6, 31-58.

————Chang, Y.S. 1976. A preliminary report on the gonadal development of adult milkfish. Chanos chanos (Forskal). In: *Proc. Internat Milkfish Workshop Conf., Tigbauan, Iloilo,* Philippines, May 19-22, 1976, 139-159.

Nash, C.E.. Kuo, C.M.. 1976. Preliminary capture, husbandry and induced breeding results with the milkfish, *Chanos chanos* (Forskal). In: *Proc. International Milkfish Workshop Conf.,* Tigbauan, Iloilo, Philippines, May 19-22, 1976, 139-159.

Chapter 7

Warm-Water Fishes in Asia and the Far East on Breeding and Selection: A Significant Review

A RESUME

This significant presentation at FAO world symposium on "Warm-water pond fish culture" at Rome in May, 1966, is that of Prof. Chaudhuri's one of the earliest work concerning the fishery development in Sixth decades of the 20th Century.

(Indian and Chinese carps, mullet and the milkfish are the most important warm-water pond fishes cultivated in Asia and the Far East. While certain carps and other fishes of minor importance spawn in ponds, the widely cultivated species do not breed in captivity.

Indian carps, however, are known to breed in special type of ponds known as 'bundhs' where riverine conditions are simulated. A better and more dependable technique has recently been evolved for inducing the breeding of these and Chinese carps by injection of fish pituitary hormones. This technique is now used in fish culture in India, China and Taiwan for raising seed of these carps commercially.

Although initial success has been achieved in spawning mullets by hormone treatment, proper breeding and hatching techniques are yet to be developed. Also, very little is known of the spawning habits of the milkfish and the eel which do not attain maturity or spawn in confined waters. Natural habitat is the only source for their seed.

> *Improvement of stock by selective breeding and hybridization is vitally important for better production of fish. Inducing sterility in certain types of fish is necessary for proper manipulation of stood in fish culture.*
>
> *The urgent need for systematic research on controlled reproduction and genetic selection of fish and also for proper training of personnel, extension and popularization of the new techniques among fish farmers is emphasized.*
>
> *To cut the presentation to size selective portions have only been reproduced eliminating elaborate report on "Breeding Techniques and selection of breeds" which deals with the species Carpinus carpio, Runtinus javanicus, Osteochihis hasselti, O sphronemess goramy, Helostoma temmincki, Tilapia mossambica, Etroplus saratensis, Puntius orphoides, Carassius carassius, Ophicephalus narulias, Ophicephalus narulins, Ophiceshalus striatus, Clarias batrachus, Heteroppenis fossilis and Anabas testudinus.*
>
> *The following presentations are excerpts of the review paper contributed by the professor which reflect his experimental successes in South and Southeast Asia).*
>
> *This presentation was Prof. Chaudhuri's one of the earliest work which was made in Rome in 1966. He gives a list of warm water species cultivated in Asia and the far East, breeding requirements, breeding habits of cultivated species, breeding techniques and selection of breeders selective breeding and hybidization and on induced breeding. In this presentation Prof. Chaudhuri stresses the development of proper breeding and hatching techniques of mullets, much more information on spawning habits of milkfish and improvement of stock by selective breeding and hybridization. He urged for systematic research on controlled reproduction and popularisation of new techniques among fish farms.*
>
> *Prof. Chaudhuri's paper on "Problems of warm water fish seed production" presented in Kyoto conference in 1976 has been briefly presented in this Chapter.*

1. Introduction

The major carps of China and India form the main economic food fishes raised in countries of Asia and the Far East. As these fishes spawn naturally in the large rivers when they are in flood, considerable difficulties are experienced by the fish farmers in obtaining sufficient quantities of pure fish seed for stocking ponds. With the first success of pond breeding of major carps in India by injections of fish pituitary hormones (Chaudhuri and Alikunhi 1957;

Chaudhuri 1960; Alikunhi *et al.*, 1960) an alternative dependable method of obtaining pure fish seed was ensured. Subsequently similar techniques were also successfully developed in artificial breeding of Chinese carps (Aliev, 1961; Alikunhi *et al.*, 1963a and b; Tang *et al.*, 1963). The Indian carps also breed in 'bundhs' which are special types of ponds subjected to flooding from adjoining extensive catchment areas (Hora *et al.*, 1945).

The breeding techniques of the other important cultivated species which spawn in ponds have been well developed. The methods adopted in different regions of South-East Asia, however, vary considerably (Hora and Pillay, 1962). Selection of proper breeders for breeding purposes is no less important since success in spawning and obtaining healthy progency depends largely on proper selection of brood fish. Selective breeding leads to improvement of stocks; the new strains or hybrids developed having better cultural qualities. Hybridization for producing hardier, disease-resistant and improved varieties of fishes (Baldwin, 1956; Scott, 1956; Buss and Wright, 1958; Chaudhuri, 1959a) has assumed increasing importance in recent years.

In the present paper an attempt has been made to bring together the information available on the methods developed in the selection and breeding of the warm-water cultivated species in Asia and the Far East.

2. List of Important Warm-water Fishes Cultivated in Asia and the Far East

The most widely cultivated fish in the region are the quick graving, non-predatory carps belonging to the family Cyprinidae. Even though the culture of piscivorous fishes is generally less economical, some predatory fishes are cultured in certain areas, mainly on account of consumer preference. In addition, a number of exotic fishes have been introduced and marine and brackish-water fishes transplanted into fresh-water ponds by the fish culturists (Jones and Sarojini, 1951, Schuster, 1952a; Alikunhi *et al.*, 1952).

A classified list of the popular pond fishes cultivated in Asia and the Far East is given below, showing the distribution and habitat of each species (Ling, 1955; Hora and Pillay, 1962).

Sl.No.	Scientific Name	Family	Distribution	Natural Habitat	Countries where Cultivated
1.	Catla catla	Cyprinidae	India, Pakistan and Burma	Rivers	India, Pakistan, Burma, Ceylon
2.	Labeo rohita	Cyprinidae	India, Pakistan and Burma	Rivers	India, Pakistan, Burma, Ceylon
3.	Cirrhina mrigala	Cyprinidae	India, Pakistan and Burma	Rivers	India, Pakistan, Burma, Ceylon
4.	Labeo calbasu	Cyprinidae	India, Pakistan and Burma	Rivers	India, Pakistan, Burma, Ceylon
5.	Hypophthalmichthys molitrix	Cyprinidae	China	Rivers	China, Japan, Hong Kong, Taiwan, Malaysia, Thailand, India, Ceylon
6.	Ctenopharyngodan idella	Cyprinidae	China	Rivers	China, Japan, Hong Kong, Taiwan, Malaysia, Thailand, India, Ceylon, Vietnam, Pakistan
7.	Aristichthys nobilis	Cyprinidae	China	Rivers	China, Japan, Hong Kong, Malaysia, Thailand, Vietnam, Taiwan
8.	Mylopharyngodon piceus	Cyprinidae	China	Ponds, rivers and lakes	China, Hong Kong, Malaysia, Thailand, Vietnam, Taiwan, Japan
9.	Cirrhina molitorella	Cyprinidae	China	Ponds, river and lakes	China, Japan, Hong Kong, Malaysia, Thailand, Vietnam, Taiwan
10.	Cyprinus carpio	Cyprinidae	Originally China and Vietnam. Now distributed all over the region	Lakes and rivers	Almost all the countries of the region
11.	Puntius javanicus	Cyprinidae	Indonesia, Thailand, Vietnan	Lakes and rivers	Indonesia, Thailand, Malaysia

Contd...

Sl.No.	Scientific Name	Family	Distribution	Natural Habitat	Countries where Cultivated
12.	*Osteochiilus hasselti*	Cyprinidae	Indonesia, Malaysia, Thailand	Lakes, rivers	Indonesia, Malaysia, Thailand
13.	*Trichogaster pectoralis*	Osphronemidae	Thailand, Cambodia, Vietnam	Rivers, shallow lakes	Thailand, Malaysia, Indonesia, Pakistan, Ceylon
14.	*Osphronemus goramy*	Osphronemidae	Indonesia, Thailand, Malaysia, Cambodia, Vietnam	Rivers, shallow lakes, swamps	Indonesia, Thailand, Ceylon, India, Malaysia
15.	*Helostoma temmineki*	Helostomidae	Thailand, Malaysia, Indonesia	Rivers, shallow lakes, swamps	Thailand, Malaysia, Indonesia, Ceylon
16.	*Tilapia mossambica*	Cichlidae	South East Africa, introduced to most of the Asian countries	Fresh-water and brackish-water river	Most of the countries in the region
17.	*Etroplus suratensis*	Cichlidae	India, Pakistan, Ceylon	Fresh-brackish water lakes	India, Ceylon
18.	*Chanos chanos*	Chanidae	Most countries of Asia and Far East	Marine and brackish-water	Philippines, Indonesia, Taiwan, India, Thailand
19.	*Mugil cephalus*	Mugilidae	Most countries of Asia and Far East	Marine and brackish-water	China, Taiwan, Hong Kong, India

Besides the above, the following species of fishes are also used for pond culture but most of these are cultured mainly in the region where they naturally occur. They are: *Cirrhina cirrhosa, Barbus (Lissochilus) hexagonolepis, Labeo fimbriatus, Carassius carassius, Parabramis pekinensis, Puntius orphoides, Mugil tade, Mugil dussumieri, Liza corsula, Pangasius* spp.*, Anguilla japonica, Ophicephalus* spp.*, Anabas testudineus, clarias batrachus, Heteropneustes fossilis* and *Lates calcarifer.*

3. Breeding Requirements and Breeding Habits of Cultivated Species

The breeding requirements of some important cultivated species are quite complex and our present knowledge of the optimum ecological conditions for their spawning is inadequate.

Fabricius (1950) studied the different stimuli for the spawning activities of fish and concluded that these were activated by an inherited neural releasing mechanism peculiar to each species in which various external as well as internal stimuli cooperate by heterogeneous summation and the release is usually brought about by several different combinations of stimuli. Of the external factors temperature and light are considered to be the most important ones influencing maturity and spawning in fishes (see Harrington, 1950; Shiraishi and Takeda, 1961; Craig-Bennett, 1930; Hoover and Hubbard, 1937; Allison, 1951). Rain, flood, increase in dissolved oxygen content and pH of water are also believed to be important external factors responsible for spawning (Khan, 1945; Mookerjee, 1945b). According to Swingle (1956) fishes probably excrete into the water a hormone-like substance or substances which may reduce or prevent reproduction. Because of this "repressive factor", certain fishes do not spawn when kept overcrowded, but spawning occurs when they are released in fresh water. Influence of population density on the reproduction of fish has been observed by Kawajiri (1949) and in Europe carp is bred by releasing brood fishes in the freshwater of "Dubisch ponds". (Schaeperclaus, 1933). Of the internal factors, endocrine secretions, especially that of the pituitary gland is believed to induce ovulation and spawning in fishes (see Pickford and Atz, 1957). This secretion is again influenced by external factors like rain, light, temperature, etc. "The external environment mediates its effect on the endocrine system through the pituitary" (Hoar, 1957).

The sexual play of brood fishes under optimum conditions also appears to stimulate secretion of these hormones.

Breeding habits differ considerably in different groups of fishes. According to their breeding habits, the important cultivated pond fishes of the region may be grouped into three distinct groups.

Group–I

This group includes fishes such as *Cyprinus carpio* (common carp), *Tilapia mossambica* (tilapia), *Osphronemus goramy* (gourami), *Trichogaster pectoralis* (sepat siam), *Etroplus suratensis* (peal–spot). *Helostoma temmincki* (kissing gourami), *Osteochilus hasselti* (nilem). *Puntius javanicus* (tawes), *P. orphoides* (mata merah), *Ophicephalus* spp. (murrels) etc. which breed in ponds.

Parental care is observed in tilapia, sepat siam, pearl-spot, gourami and the murrels. Tilapia (male) makes a circular depression at the pond bottom in which the female lays eggs. The fertilized eggs are picked up by the female and kept in the mouth where further development and hatching take place (Vaas and Hofstede, 1952). The gourami builds a nest out of grass and weeds on the pond margins, while the male sepat siam builds a foam nest in which the eggs are laid. The pearl-spot attaches eggs in layers to hard substrata. The murrels build nests with vegetation near the margin of ponds in shallow waters and the parent fish guard the fry.

Other cultivated species belonging to this category such as the common carp, nilem, tawes, mata merah and kissing gourami do not show parental care but breed in ordinary ponds. Nilem and tawes spawn when well-oxygenated running water is supplied to the ponds. While the eggs of nilem and tawes are demersal and non-adhesive, those of kissing gourami float on the surface and the mata merah eggs are adhesive. The common carp also breeds in ordinary ponds but the intensity of breeding is accelerated when they are released in fresh water. The adhesive eggs are attached to aquatic plants or on other objects. Majority of the fish belonging to this group breed more than once or several times in a year.

Group–II

Fishes of this category do not ordinarily spawn in ponds but breed in flooded rivers and adjoining areas. Indian and Chinese carps belong to this group- Indian carps are known to breed in 'bundhs' which are specially designed ponds receiving inflow of fresh rain water (Das, 1917), but Chinese carps were believed to breed only in their natural habitats in China. Kuronuma (1954) reported the breeding of silver and grass carps in the Tone river of Japan. Tang (1960) has also obtained fry and fingerlings of these carps in the Ah Kung Tian reservoir in Taiwan.

However, in recent years, both these Indian and Chinese carps could be induced to breed.by injection of fish pituitary hormones (Chaudhuri and Alikunhi, 1957; Chaudhuri 1960; Aliev, 1961; Alikunhi *et al.*, 1963a and b; Tang *et al.*, 1963) as already mentioned and the technique has been further developed to obtain commercial production of fish seed (Chaudhuri, 1963; Tang, 1965; Chaudhuri *et al.*, 1966).

Group–III

This group includes mainly marine or brackish-water species, which do not spawn in ponds or so far could not be induced to spawn in ponds. *Chanos chanos* (milkfish), *Lates calcarifer* (cock-up), *Mugil* spp. (mullets) and *Anguilla japonica* (eel) come under this group. The milk-fish breeds in the sea and the eggs are pelagic. The species has not been observed so far to attain maturity or to spawn in ponds or in other confined waters. The grey mullet (*Mugil cephalus*) also does not breed in ponds or in estuaries and is believed to spawn only in the sea. The fresh-water mullet (*Liza corsula*) breeds in the fresh waters of rivers (Pakrasi and Alikunhi, 1953) and the cock-up in brackish-water rivers, creeks or lagoons. The eel spawns only in the deep sea. The young ones of most of the above mentioned species are collected from the sea coast and estuaries for stocking in ponds.

4. Induced Breeding by Hormones Injections

The author dealt with the breeding of Indian carps in special type of ponds called 'bundhs'. The technique, however, has its limitations and cannot be universally practiced since the 'bundhs' can be constructed only in areas where the topography is one of undulating terrain. Moreover, the construction and maintenance of a perennial or seasonal bundh is very expensive and beyond the means of average fish farmers. Lastly, since bundh breeding is dependable to a great extent on the climatic and environmental conditions, success is unpredictable and poor results are very after reported due to failure of rains. Hence, the need to develop better and more dependable source of fish seed has been immensely felt. The recent success in breeding the above mentioned carps by administration of fish pituitary hormone has opened new possibilities in solving the pressing problems of dearth of seed of these economic food fishes.

5.   History of Pituitary Treatment in Fish Culture

Before going to the details of the technique followed in inducing breeding of carps, it is desirable to make a passing reference to the history of pituitary treatment in pisciculture. Atz and Pickford (1959) have given an account of the use of pituitary hormone in fish culture in different countries of the world.

Brazil was the first country ever to develop a technique of breeding fish by injection of fish pituitary hormones. Following the work of Houssay (1930) of Argentina, von Ihering and others in Brazil started experimenting with various hormone injections and finally succeeded in breeding Brazilian fishes by the administration of fish pituitary hormones in the year 1943 and obtained normal young ones (von Ihering and Wright, 1935; von Ihering, 1937). Since then the Brazilian fish culturists are using the technique to obtain seed of the indigenous fishes as part of their regular piscicultural program. Fontenele (1955) has given a detailed account of the procedure followed by the Servico de Piscicltura in Brazil.

The Russians were the next to introduce hormone treatment in fish culture. They have made extensive use of this technique for solving the problems caused by the construction of dams in many rivers in which sturgeons and other important migratory species breed. With the first success in breeding sturgeons in 1937 by Gerbilskii (1938), the pituitary treatment method is being successfully applied in fish culture for obtaining all eggs of sturgeon in the sturgeon farms situated on the lower Volga, Ural, Kura, Kuban and other rivers (Gerbilskii, 1965).

In the United States, the technique is mainly employed in the production of bait minnows (Ball and Bacon, 1954) and for the culture of channel catfish, *Ictalurus punctatus*, a commercial as well as sport fish (Clemens and Sneed, 1957; Riggs, 1958). Attempts have also been made to induce spawning in Pacific salmon (*Oncorhynshus* spp.) which are prevented from reaching their spawning grounds by the construction of dams in the rivers but success has not been achieved so far (Burrows *et al.*, 1952; Palmer *et al.*, 1954).

In India, the first success in inducing breeding of major carps by pituitary injection was achieved in 1957 (Chaudhuri and Alikunhi, 1957) and since then the technique has been used as a regular part of fish cultural procedure. More recently, a similar technique has been successfully employed in the breeding of the Chinese carps also (Alikunhi *et al.*, 1963b; 1965).

In Taiwan, fish culture has been rejuvenated by the success obtained recently in spawning the Chinese carps by pituitary treatment (Tang *et al.*, 1963) and promising development towards commercial production of fish has been initiated (Tang, 1965; Lin, 1965).

Fukushima (1965) has reported successful spawning of Chinese carps by hormone injections, throughout the mainland of China and a production of 1200 million fry was reported to have been achieved during 1962.

6. Breeding of Indian Carps

The Indian major and minor carps such as *Catla catla, Labeo rohita, L. calbasu, L. gonius, L. fimbriatus, L. bata., L. boga, Cirrhina mrigala, C. reba* etc. are the important pond fishes successfully bred in India by injection of fish pituitary hormones. The pituitary glands for injection are collected just prior to or at the beginning of the breeding season from fully mature gravid fish. Glands from immature or spent fish do not give satisfactory results. However, glands of induced bred fish, collected immediately after spawning have been observed to be potent and are effectively used. Ordinarily glands from the same species as the recipient fish are used, but glands from other closely related species as also from fishes belonging to different families have given positive results. Care is taken to collect glands from fishes soon after capture, although those procured from fishes preserved in ice are also found to be potent.

Preservation of glands is made in absolute alcohol and preferably stored under refrigeration in air-tight glass stoppered phials. The glands after proper preservation are weighed individually or collectively in an electric balance. The weight is noted exactly' after two minutes calculated from the time of removal of the gland from absolute alcohol. The weighed glands are again transferred to phials in absolute alcohol and stored till the time of injection, when the required quantity of glands as calculated from the weights of the breeders to be injected are taken out, allowed to dry for a few seconds and then macerated in a tissue homogenizer with a little distilled water or 0.3 per cent common salt solution. The homogenized glands are next diluted with the same liquid, the rate of dilution being 0.2 ml per kg body weight of the breeders. The gland suspension is next centrifuged and the supernatant fluid is taken in a hypodermic syringe for injection. To avoid preparation of the extract every time just prior to injection, a simplified method has been successfully tried by preserving the extract in glycerine and then storing either in an air-tight phial or preferably ampuled and stored for future use (Ibrahim and Chaudhuri, 1965).

Determination of proper dosage depends mainly on the proper stage of sexual maturity of the breeders. By careful selection of breeders and injecting an adequate weight of pituitary gland per kg body weight of the breeders 60-100 per cent success can be obtained provided the water temperature is optimum. Depending on the condition of the female breeder, a single injection of 5 to 10 mg of homoplastic pituitary gland per kg of its body weight may give successful spawning. In usual practice however, the female alone is given a preliminary dose of 2 to 3 mg per kg body weight and kept segregated. After 6 hours a second dose of 5 to 8 mg per kg body weight is administered to the female and a first dose of 2 to 3 mg per kg body weight is given to the males, after which the males and female are put together for breeding. Intramuscular injections are given on the caudal peduncle or near the shoulder region using a 2 ml hypodermic syringe graduated to 0.1 ml division. B.D. needle No.22 is used for fish weighing 1.0 to 3.0 kg and No.19 is used for larger ones. For smaller fishes needle No.24 can be used.

The major carp breeders are raised in ponds. A few months prior to the breeding season, 2 to 4 years old adults are collected and stocked in brood fish ponds for breeding purposes. Proper care of the breeders is taken and with the advent of monsoon when fishes become fully' mature and the temperature of water goes down, breeding operations are started. A male breeder can be easily distinguished by the roughness of the dorsal surface of the pectoral fin in contrast to that of a female which is 'smooth to touch (Chaudhuri, 1959a). Fully ripe males which freely ooze milt when gently pressed on the abdomen are selected. Selection of a female breeder in prime condition is rather difficult. Ordinarily a female having a soft, bulging rounded abdomen and with swollen, reddish vent is selected for breeding. Female breeders are also cathetered before selection so as to find out the condition of eggs. Healthy breeders weighing 1.5 to 5.0 kg are preferable, as they are easy to handle.

Selected breeders are collected in hand nets and weighed. The quantity of pituitary glands required is next calculated from the weights of breeders and the suspension prepared. The breeder wrapped inside the hand net is then placed on a soft cushion and injected intra- muscularly. One set of breeders consisting usually of two males and one female is then introduced into one breeding hapa. The breeding hapa are fixed by bamboo poles in the marginal waters of ponds.

Spawning ordinarily takes place within 3 to 6 hours after the release of the injected breeders. In case no spawning occurs, a third injection with a slightly higher dose is given 10 to 12 hours after the second injection to the females. No vigorous sex-play is observed as in nature but chasing of the female by the male with a little splashing of water is usually noticed at the time of spawning. The eggs are non-adhesive, demersal and swell up considerably after fertilization to the size of a pea. Approximately 8 to 10 hours after fertilization, the embryo starts twitching movement. By that time the eggs get properly water hardened and are removed from the breeding hapa. Breeders are then collected and usually sacrificed for collecting glands.

A quantitative assessment of the eggs are made from the total volume and percentage of developing eggs are estimated from samples. The eggs are next distributed uniformly to a number of hatching hapa.

The hatching hapa are tied to bamboo poles in marginal waters of ordinary ponds. About 0.075 to 0.1 million eggs are uniformly spread on the stretched bottom of the inner hapa. The eggs hatch out in 15 to 18 hours after fertilization at a temperature of 21 to 31^0C. After hatching, the hatchlings escape to the outer hapa through the meshes of the inner one. Egg cases and bad eggs are left in the inner hapa which is removed when all the eggs have hatched out. The hatchlings are left undisturbed in the outer hapa till the third day after hatching. By that time yolk gets almost absorbed and the young fry start feeding. They are collected and stocked in nursery ponds.

The hydrological and climatic conditions studied (Chaudhuri, 1960) during induced spawning of carps indicated that–(*i*) spawning occurs both in clear as well as in turbid waters and during day or night; (*ii*) injected fish breed successfully when the water temperature ranges between 24 and 31°C, the optimum being around 21°C. Even if the fish spawn at a higher temperature, the percentage of fertilization and hatching is relatively low; (*iii*) fish breed at a fairly wide range of pH and dissolved oxygen contents of water; (*iv*) cool, rainy days give better results than hot, sultry days; and (*v*) rain and fresh rain water are conducive to successful spawning.

The technique of breeding of Indian carps as described above is more or less followed throughout India with slight modifications suited for the particular environment. In several places where the water temperature in ponds is above the optimum the injected carps are introduced in breeding hapa fixed in flowing waters in rivers. The temperature of river water is at least 2°C below the pond temperature. At the Central Inland Fisheries Research Sub-station, Cuttack, preliminary attempts made to breed carps in the air-conditioned laboratory at controlled temperature (21 to 28°C) gave encouraging results. One set of rohu and one set of calbasu were bred successfully in 1962 and subsequently Alikunhi *et al.* (1964a) conducted elaborate experiments and obtained 0.8 million spawn.

7. Breeding of Chinese Carps

In India

The Chinese silver and grass carps were first introduced into India when experimental consignments were brought to Cuttack during September and December, 1959, from Tokyo and Hong Kong respectively. Both the species attained sexual maturity at the Central Inland Fisheries Research Sub-station ponds at Cuttack, when about two years old (Alikunhi and Sukumaran, 1964). During July 1962, when the fish were just three years old, both the species were successfully induced to breed in ponds by injection of fish pituitary hormones. An injected bighead (*Aristichthys nobilis*) also ovulated by hormone injections (Alikunhi *et al.*, 1963a; 1963b). Mature males of these Chinese carps are easily distinguished by the processes and ridges on the pectoral fin rays (Tang, 1954). The techniques of breeding adopted were essentially the same as those followed in the case of Indian carps. Success was achieved with homoplastic as well as heteroplastic injections of pituitary glands collected from Indian carps. A total of 9 mg per kg body weight injected in two doses to females gave successful results. The female usually starts oozing 6 to 8 hours after injection. While in Indian carps natural spawning occurs after injections, the Chinese carps generally have to be stripped and the ova artificially fertilized. Chaudhuri *et al.* (1964), however, has observed natural spawning in breeding hapa of injected grass carp and subsequently also of silver carp. Alikunhi *et al.* (1965) found the progeny of induced-bred silver carp attaining maturity in the first year itself. The yearlings could be bred and young ones obtained, but in a large number of instances the hatchlings did not survive. In these experiments too, spawning occurred 4 to 8 hours after the second injection to the female but the eggs laid in the breeding were not fertilized and artificial fertilization method had to be adopted. In grass carp yearlings, however, only the males were ripe and oozing. Two year old grass carp have been successfully bred (Chaudhuri *et al.*, 1966).

In Taiwan

It was during 1963 fish breeding season that first success was reported in inducing spawning of the Chinese bighead, silver and grass carps in Taiwan (Tang *et al.*, 1963). The method followed is briefly described below.

Breeders were collected from a reservoir and released before injection in holding tanks with continuous circulation of water. Carp pituitary glands preserved in absolute alcohol were extracted in physiological solution and injected intramuscularly in mature spawners. Females with dilated genital papilla and flaccid abdomen and males with oozing milt were selected for breeding. Successful spawning was obtained in a few which ovulated in 6 to 18 hours after receiving a single injection of one whole gland collected from a carp of comparable size or after receiving a second or third injection of the same dosage. Many did not respond even after repeated injections. The males received a lesser dose of 0.5 to 1.0 pituitary gland. Artificial fertilization of eggs with sperm by hand stripping was followed in all the successful experiments with the single exception of one bighead which was found to ovulate spontaneously in a pond 18 hours after receiving a single injection. The dry method of artificial fertilization was adopted.

About 33 per cent success was achieved. The hatching rate in a solitary case was as high as 95 per cent but in the majority of cases only about 20 per cent or less hatching was observed. The reason for this was mainly attributed to overdose of hormone injected. Lowering of the temperature was believed to be one of the most important environmental factors conducive to induced spawning.

Within a year of the initial success improvements were effected in the breeding and hatching techniques employed for inducing spawning of the Chinese carps and commercial production of their fry was started in a large number of hatcheries (Tang, 1965).

Experiments with fish pituitary combined with chorionic gonadotrophin gave more effective results than administering fish pituitary alone. With this improved method the percentage of success was raised from 33 to 78, thereby achieving an increase of 45 per cent success. A number of brands of gonadotrophin were tried, of which Puberogen (contains 250 IU/CC of CG), Gonagen and Synahorin (both contain CG mixed with a little of mammalian pituitary extract) gave encouraging results.

A circular basin with a steady stream of circulating water, which eventually flows out through an opening in the centre (Tang, 1965), was found to be very efficient for hatching.

Lin (1964) as cited by Tang (1965), succeeded in spawning the mud carp, *Cirrhina molitorella* and the technique followed was almost identical to that of other Chinese carps. He succeeded by injecting glands from the skipjack tuna. Positive results were also obtained by injecting 500 rabbit units of cattle anterior lobe extract combined with fresh toad pituitaries. In all cases of breeding of these Chinese carps, artificial fertilization of hand-stripped eggs was adopted.

Liu (1964) has succeeded in breeding silver carp as early as two months before the normal breeding season. This prolongs the growing season of fish and is a definite advantage to the fish farmers.

In Malaysia

Attempts in inducing the breeding of grass carp by hormone treatment were made in Malaysia but did not meet with much success. Hickling (Personal communication) obtained ovulation in grass carp a number of times by injection of pituitary hormones at the Tropical Fish Culture Research Institute, Malacca, but the eggs could not be fertilized. Artificial fertilization of eggs was tried by him by both the wet and dry methods but no positive results were obtained. He is of opinion that probably the absence of a definite seasonal change of daylight and temperature at Malacca is responsible for the improper maturation of gonads and it is likely that the final meiotic division in the ovum does not take place.

In Japan

Natural propagation of Chinese carps have been observed in recent years and all the species are reported to have established in the Tone river system in Japan. Japanese workers have also succeeded in obtaining large numbers of these species by hatching eggs collected from these rivers (Kuronuma, 1958; Inaba *et al.*, 1958; Suzuki *et al.*, 1958; Suzuki and Takahash1, 1961).

Kawamoto (1950) tried to obtain spawning in grass carp in Japan by feeding the fish regularly with sex hormones mixed with the regular diet. The fish did not spawn, but the application of sex hormones definitely hastened the development of gonads. Injections of hormones has also been tried on Chinese carps and on several other marine and fresh-water fishes. Although no published accounts of these have come to the knowledge of the author. It is understood that the treatment with synahorin in combination with fish pituitary hormones has been tried with success by a few fish farmers.

In China

There is hardly any detailed information or published account available on the induced breeding work carried out in the mainland of China. Isaev (1958) has published an account of the fish breeding in the Chinese People's Republic (not consulted by the author). Recently some information on this subject has been gathered by Fukushima (1965) who reports that all important Chinese carps are bred by hormone injection allover the mainland of China and that fish seed is produced on a large scale by this method. Chorionic gonadotrophin is used to obtain eggs from black carp and big head.

8.   Breeding of Catfish

Majority of the cultivated species of catfishes breed in ponds. Induced breeding by pituitary treatment could be suitably tried in a few species which do not breed in ponds or for hastening pond breeding.

The large-sized catfish Wallago attu which is common in ponds and in larger impoundments in India and Pakistan, breeds in rivers. This fish has been reported to spawn in bundhs, but so far no attempt has been made to breed the species by hormone injection. The smaller species *Ompok bimaculatus* (Bloch) which is primarily a riverine fish and is cultivated in ponds, has been induced to breed in cisterns by the injection of 0.4 to 0.5 of a pituitary gland collected from the carps rohu and mrigal (Chaudhuri, 1962).

The Siamese catfishes *Pangasius larnaudi* Bocourt and *P. sutchi* Fowler, also cultured in Vietnam, do not breed in ponds and their fry are collected from rivers. Large-scale breeding of these fishes by hormone treatment has not yet succeeded although Tubb (1958) has reported early spawning of *P. sutchi* accomplished by hormone injection. Two other cultivated species, *Clarias macrocephalus* Gunther and *Clarias batrachus* readily spawn in ponds and especially the latter species can be bred by the simple replacement of old water by fresh rain water. As such, there is no necessity for hormone treatment in these species for breeding. But in certain cases the pituitary extract is used for getting early spawning of these fishes. Tongsanga *et al.* (1963) achieved early spawning in mature catfish *C. masrocephalus* by giving injections of 13 mg, 26 mg and 39 mg of homoplastic pituitary glands per fish 90 to 190 g in weight. The higher doses of 26 mg and 39 mg gave cent per cent success.

9.   Breeding of Mullets

Excepting *Liza corsula*, which was observed to mature in Pulta Water Works settling tanks near Barrackpore, West Bengal, there is no report of other mullet species attaining full maturity in fresh-water ponds. Pakrasi and Alikunhi (1952) observed breeding of *Liza corsula* in fresh-water rivers during June and July and states that the fish appears capable of breeding both in brackish as well as in fresh-water habitats and the breeding season seems to be a prolonged one extending from May to September.

Mugil cephalus is the most important cultivated mullet and has a wide distribution. The only means of obtaining seed of this species is by collecting it from the sea coast, creeks, lagoons and adjoining areas. Besides the difficulties of collection and transport involved, the fry collected are usually a mixture of economic and uneconomic varieties. Attempts were therefore, made in a few countries to induce them to breed by injection of hormones and raise the fry artificially.

Preliminary attempts were made in India to breed the two important economic species of Chilka lake, namely *Mugil cephalus* and *Liza troschelii*, by hormone treatment. (Proc. Indo-Pac. Fish Coun., Section 1:p 40, 70, 1963). Experiments were conducted in lake water and breeders were collected from the sea coast near the lake mouth. Injection of homoplastic pituitary gland extracts (8 mg to 16 mg per kg body weight of breeders) resulted in ovulation of eggs on several occasions in both the species. Both the species could be stripped successfully, but the eggs failed to develop in lake water. When sea water was used, normal development of eggs proceeded on several occasions and on one occasion a *Mugil cephalus* egg hatched out in 48 hours (22.5 to 23.5°C) after fertilization.

The low salinity of lake water was found to be responsible for failure in getting fertilization of eggs in lake water. It was observed that *M. cephalus* sperms died immediately when brought in contact with lake water. *L. troschelii* sperms, however, were active for about a minute. On the contrary, sperms of both the species were found to be active for about ten minutes when mixed with sea water.

The results indicated that it may be possible to get successful spawning of these mullets if experiments were conducted in sea water with facilities for water circulation which appear essential for proper development of eggs and survival of hatchlings.

This work was continued and success has been achieved by the Orissa State Fisheries staff (Personal communication) in 1963 in artificially fertilizing the stripped eggs of injected *L. troschelii* in sea water and 10,000 spawn were produced which survived for over a week.

Tang and his co-workers in Taiwan carried out experiments during 1963 mullet fishing season and successfully bred *Mugil cephalus* by hormone injections (Tang, 1964). They were collected from the sea during their spawning migration to the south-western coast of Taiwan. Brood fish were held in special boxes in shallow sea water. The breeding technique followed was identical to that used for breeding Chinese carps. Mullet pituitary extract in combination with Synahorin (a mixture or CG and mammalian pituitary extract) was used. No ovulation occurred when 1.0 to 3.0 pituitaries alone were injected or when the fish received combined extract of less than 2.0 pituitaries plus 40 rabbit units of Synahorin. Positive results were obtained when 2.0 fish pituitaries combined with 40 rabbit units of Synahorin were injected, which is considered to be the threshold dosage for successful spawning. Artificial fertilization of hand-stripped eggs was made following the dry method. The developing eggs were hatched in aquaria with circulation of sea water and also under natural conditions near the sea coast. Hatching took place in 59 to 64 hours at a temperature 20.0 to 24.5°C and in salinities of 24.39 to 35.29 parts per thousand. Most of the larvae died within two days after hatching and none survived beyond four days.

High mortality of brood fish after injection hampered the experiments. The authors believe that the response of these mullets to hormone injections could be greatly improved if the fish were kept in better condition at the time of the experiment.

The preliminary success in India and Taiwan, as mentioned above, will go a long way to developing a dependable technique for the spawning and successful rearing of the larvae and early fry of mullets.

10. General Discussion on Hormones Treatment in Fish

Successful hormone treatment of fish and its use in pisciculture started in Brazil in the early thirties when fish pituitary hormones were first used for inducing spawning in fish. This success appears to have given a fresh stimulus to physiologists to conduct intensive research on the physiology of the pituitary gland and its effect on the reproductive behaviour of fishes. Pickford and Atz (1957) in their book on "The physiology of pituitary gland of fishes" have summarized all available information and discussed various aspects of the physiology of fish pituitary gland and its role in present day fish culture.

Although successful experiments have been conducted and techniques developed in a number of countries to breed fish by hormone injection, bioassay methods have not been developed in any country other than Russia. Atz (Pickford and Atz, 1957) states "Chemical standardization of fish gonadotrophins is of course a thing of the future, and the Russians are the only ones who have developed a method of biological assay". Although the Russians have found out a "Vy'un Unit" (Iazanski, 1949) for assaying fish gonadotrophin, they have yet to solve the problem of dosage of hormone. Gerbilskii (1965) has given an account of the technique of hormone treatment in fish culture followed in Russia. He states that the effectiveness of pituitary injections depends mainly on the coincidence of the time of treatment with the suitable period of the sexual cycle of fish. The study of fish ovogenesis is of great importance to "ensure effective application of pituitary injections and to forecast precisely enough the moment of mating". Sample of ovocytes are collected from injected sturgeon female every five hours during the period of maturing until the completion of ovulation. By making a thorough study of the complex processes, Deltaf and her collaborators, as cited by Gerbilskii (1965), have been able to recommend the proper time for obtaining eggs from the female. In Brazil, the techniques have not been standardized. In the absence of any accurate unit of measurement, Fontenele (1955) suggested a conventional unit which is the amount of hormone obtained from a whole pituitary gland from a Curimata of well-developed gonads. Recently Clemens and Sneed (1962) attempted evaluation of the activity and relative potency of pituitaries of various species of fish collected during various months of the year and also determination of phylogenetic specificity, if any, in pituitary glands by bioassay methods. In India, however, although the pituitary hormones were not isolated and chemically standardized, fairly satisfactory results were obtained by precautions taken in selection of proper recipient fish (breeder) as well as the donor fish and also in preservation and storage of glands. The dosage given has been always in mg weight of the pituitary gland per kg body weight of fish (Chaudhuri, 1963). Female breeders are also cathetered and eggs examined before final selection.

In general, pituitary glands from fully mature gravid fish are used in induced fish breeding experiments, which are believed to be the most potent. Gerbilskii (1940) observed relatively low potency of gonadotrophins in the pituitaries of spent fish and the same was observed by de Menezes (1945) in the case of immature fish. But Clemens and Sneed (1962) observe "Activity in relation to size, sex and degree of gonadal development of the donor fish does not appear as important as implied by some previous reports in the literature". As regards relative potency of the male and female pituitary glands, the results of research conducted in Brazil (Fontenele, 1955) and also at the Central Inland Fisheries Research Sub-station, Cuttack, India, have shown that the pituitary glands from male and female fish are equally potent. Ramaswamy and Sundararaj (1957), however, observed that in the catfish *Heteropneustes fossilis*, the pituitary gland of 'the male is not so potent as the female during the breeding season. The results obtained by the said authors, however, are not strictly comparable as the relative potency was determined by injection of the whole or part of a pituitary and not according to the weight of gland or by any other bio-assay method. Besides, there is difference in size of the donor catfish as males are usually smaller than the females.

In preserving glands, both acetone drying and preservation with absolute alcohol have been found to be effective and are popular with the fish culturists allover the world. Neto and Rebeiro (1946) prepared fish pituitary extracts in glycerine and found that the solution retained its effectiveness for a year when kept sealed in air-tight vials at room temperature. According to de Menezes *et al.* (1945), glycerine preserved pituitary extracts are more uniform in potency than those in saline or distilled water. Ibrahim and Chaudhuri (1965) found glycerine method very effective and more convenient. Clemens and Sneed (1962), however, are of the opinion that although glycerine extracts are easier to inject, their preparation is tedious. They also observed that administration of more than 10 per cent glycerine has harmful effect on the recipient fish.

Phylogenetic specificity in fish pituitary hormones is a subject of controversy. Creaser and Gorbman (1936) believe that there is a specificity in hormone physiology and a qualitative species variation exists in gonadotrophic hormones. When the variation is sufficiently great between widely separated donor and recipient species, it leads to apparent ineffectiveness of the hormone. This view is also supported to some extent by Hoar (1957) who, while reviewing the work of Witschi (1955) pointed out that the refractoriness of fish to mammalian gonadotrophins is probably due to different biochemical materials operating in groups as widely separated as mammals and fish. According to Pickford and Atz (1957), phylogenetic specificity does not appear to be significant in fish culture. Clemens and Sneed (1962) also observed that the specificity is so slight that it has no practical bearing in fish cultural practices. Chaudhuri (1960) reported successful spawning with heteroplastic pituitaries from fishes of the same family as well as from closely related families. Ramaswami and Sundararaj (1957) succeeded in spawning gravid catfish with homoplastic glands or with glands from genera of the same order but the fish became refractory when injected with pituitaries from fish belonging to different orders. However, the same authors reported successful spawning in the catfish by using frog pituitary hormones. tin (1964), as cited by Tang (1965), succeeded in breeding the Chinese mud carp by injecting pituitaries from skipjack tuna, belonging to a different order not closely related to the carps.

These and other results achieved by a host of earlier workers experimenting with mammalian pituitary hormones, sex hormones and various sex-steroids as reviewed by Pickford and Atz (1957) and Chaudhuri (1956), demonstrated the relative effectiveness of fish pituitary hormones on the spawning of fish, over those of mammalian origin. The fish pituitary hormone was, therefore, used in all the countries for spawning fish as part of regular fish culture work.

Recent findings, however, show that the mammalian pituitary hormones in combination with the threshold dose of fish pituitary glands are able to precipitate spawning in fish (Ramaswamy and Lakshman 1959; Tang, 1965). But out of all hormones of mammalian origin, the chorionic gonadotrophin has been found to be the most effective on fish. Hoar (1957) states "pregnancy urine and chorionic gonadotrophins frequently give positive results but it is difficult to say which of the many factors present on these materials is active". According to Burrows (1949), the chorionic gonadotrophins (CG) behave primarily as a luteinizing hormone (LH) and since LH is believed to be responsible for the growth and maturation of gonads in fish (Kirshenblat, 1956), the effectiveness of CG on fish can probably be explained. Sneed and Clemens as cited by Riggs and Sneed (1959) induced spawning in the channel catfish and obtained large-scale production of fry by administering human chorionic gonadotrophin (HCG). Ramaswamy and Lakshman (1959), injected a pond catfish with 250 I.U. of chorionic gonadotrophin (known by the trade name of 'Physex') and obtained ripening of eggs. But considering the high cost as compared to fish pituitary hormones, it was not tried in fish culture practices in India. In Russia, as early as in 1947, HCG was used to diagnose pregnancy I' in women with the fish loach, as the test animal (Gerbilskii, 1965), but HCG was not immediately used in their regular breeding of sturgeons. With the increase in the demand for pituitaries in recent years and difficulties in their mass procurement, Russian workers felt the need of a substitute which could be produced commercially.'Choriogonin' (a trade name for CG) was first used in 1964 on loach with great success. Among other possible substitutes, several preparations such as 'Szhk', prepared from pregnant mare serum, 'Estrovet' and 'Hypophysin Forte' were tried by the Bulgarian fish culturists (Bratanov *et al.*, 1963) on carp and trouts (as cited by Gerbilskii, 1965). The authors concluded that these preparations reduce the duration of the spawning process.

Tang (1965), while summarizing the work on induced breeding of Chinese carps in Taiwan, states that treatment with fish pituitaries in combination with chorionic gonadotrophin increases the effectiveness and better success is achieved. Synahorin (CG mixed with certain amount of mammalian pituitary extract) was used with fish pituitaries and 78 per cent positive response was obtained. He, however, does not mention the dosage and no systematic experiments were carried out with proper controls to arrive at the conclusion. Fish culturists are using different dosages of synahorin and fish pituitary glands and each hatchery has its own method.

Results of bioassay experiments conducted in recent years by Ramaswamy and Lakshman (1960) with trichloroacetic acid (TCA) and a few other chemicals on the glycoprotein content of catfish pituitary gland showed that both the glands immersed for 6 or 12 hours in 1.25 per cent TCA and the acid fluid in which they were immersed brought about spawning, but the glyco-proteins are extracted and denatured in 12 to 24 hours. TCA of 2.5 per cent strength however did not give the desired results. TCA extract of pituitary glands was used at the Central Inland Fisheries Research Institute in India for breeding Indian major carps. Pituitary glands immersed for six hours in 1.5 per cent TCA were successful in spawning in II out of 12 fish. The minimum effective dose was 14 mg per kg of the fish. Three sets of rohu, when injected with 2.5 per cent TCA in which glands were immersed for six hours, did not respond, but when the duration of immersing was reduced to three hours, another three sets could be spawned.

From the foregoing discussion on the various attempts made to spawn fish by hormone treatment, it will be seen that a number of techniques are followed by different workers which produce varying degrees of success. The fish breeder should be able to select the best method suitable for the particular species of fish he breeds. There is, however, ample scope for improvement of the breeding technique by the use of proper bio-assay methods and by chemical standardization of the gonadotrophins.

11. Selected Breeding and Hybridization

Carp culture has been a very old practice, especially in India and China. Although the Chinese and Indian carps have been cultured for thousands of years they could not be properly domesticated and they do not differ much in characters from their wild ancestors. The common carp, however, has been domesticated and long-time selection has produced a large number of strains and varieties. This has been possible only because of the fact that unlike the Chinese and Indian carps, the reproduction in common carp can be controlled. The basic need of fish culturists to have better fish for increased production, makes them select the quickest growing, best shaped and healthiest fish as breeders. Such selection finally results in overall improvement of the stock.

Hybridization is another method of combining desirable qualities and of raising hybrids superior to the parent species. In recent years, study of hybridization in fish has assumed increasing importance in both theoretical and applied ichthyology. In the West, systematic hybridization of various species of salmon and trout has been done (Scott, 1956; Baldwin, 1956; Donaldson *et al.*, 1957). In the U.S.S.R., considerable importance is attached to genetics and hybridization work in fisheries. Several interracial crosses of sturgeons and both inter-specific and inter-generic hybrids have been produced there (Gerbilskii, 1952; Kazanski, 1953). Hybridization in Chinese and Indian carps, however, could not be done till success in artificial breeding of these fishes was achieved.

Chaudhuri (1959b; 1961) first succeeded in producing hybrids in Indian major carps by artificial fertilization of hand-stripped eggs. He obtained 12 different hybrids by crossing carps comprising three genera and six species. Out of these, nine hybrid species were reared to adult size. At least three of the above hybrids showed superior qualities over the parent species. Alikunhi and Chaudhuri (1959) produced a hybrid by fertilizing eggs of an Indian carp (rohu) with common carp sperms and a few specimens grew to adult size. Subsequently Alikunhi *et al.,* (1963a; 1963b) succeeded in the hybridization of Chinese carp species and" also of Chinese and Indian carps but none of these survived beyond one week. Tang (1965) has reported successful hybridization between bighead and silver carp. These results demonstrate clearly that the hormone injection method has opened a new line of research in selective breeding and hybridization of these economic food fishes of the Indo-Pacific region, with a view to developing improved strains.

12. Induced Sterility

Sterility may be induced in cultivated fish either by injecting hormones or by hybridization. Control of population density in ponds becomes possible by stocking sterile hybrids. This has far reaching value in scientific fish culture since proper manipulation of stock is essential for obtaining higher production in a pond.

According to Hickling (1962), "Inter-specific hybrids, or mules are usually sterile". This may not be true with carps. At the Central Inland Fisheries Research Institute in India an inter-specific hybrid (cross between male *Labeo rohita* and female *Labeo calbasu*) was bred by hormone injections and produced 40,000 fry of F_2 generation of the hybrids. Even an inter-generic hybrid (cross between *Catla catla* male and *Labeo rohita* female) was fertile and could be successfully induced to spawn by hormone injections and thousands of F_2 generation hybrid fingerlings were obtained. Hubbs (1955) however is of the opinion that the great majority of hybrids in nature as well as in captivity are sterile. Such hybrids according to him are predominantly males having abnormal testes and spermatozoa. Hickling's experiment (1960) with Tilapia mossambica demonstrated that when an African male 1 was crossed with a Malayan female the offspring were all males, or nearly so. Although these male hybrids do not show any sign of being either sterile or intersexes, production of all male offspring has considerable significance in fish culture. As already discussed in previous chapters, such hybrids could be profitably used for mono-sex culture of tilapia. Recently, Suzukii (1961) from Japan has reported sterility in artificially produced inter-generic hybrids among bitterlings.

13. Suggestions for Future Work

Success in induced breeding of the Indian and Chinese carps, by administration of hormones, has greatly helped in large-scale production of pure fish seed, but this is just the beginning and there is vast scope for further improvements. Intensive research directed towards chemical standardization of hormones by proper bio-assay methods has to be carried out and useful data on the sexual cycles, gametogenesis and control of these processes in cultivated food fishes have to be collected in order to perfect the induced fish breeding technique and to improve hatching methods further.

Since hormone treatment in fish culture mainly brings about transition of the fish from fourth stage of maturity to the fifth one, the technique is successfully applied only to fish already close to natural spawning. It can reasonably be hoped that many of the present difficult-to-spawn fish which either attain full maturity in confined waters or the fully mature specimens of which can be collected from their natural habitat, can be bred by hormone injections, provided proper facilities are available. The technique has not succeeded in fish whose gonadial condition does not reach the fourth stage in ponds. Even if prolonged hormonal treatment is effective in obtaining the sexual maturity of such fish, it will be difficult to adopt it in commercial fish culture. Further research should, therefore, be directed towards improvement of environmental conditions favouring maturation of gonads or evolving the right type of artificial or natural food which would bring the fish to maturity.

Development of strains of cultivated fish which would be more efficient in food conversion, more resistant to diseases and unfavourable environmental conditions, have quicker growth rates, higher fecundity and earlier maturity, is of vital importance in fish culture in Asia and the Far East. Intensive research work to achieve this is urgently required in the area.

14. Problems of Warm-Water Fish Seed Production

(Paper Presented in 1976 Kyoto conference by H. Chaudhuri and S.P. Tripathy)

Many species of warm water fish spawn in rivers on estuaries or in the sea making it difficult to obtain their seeds for culture operation. For certain species artifical methods of propagation have, however, been developed for some species. Yet, the Techniques of successful spawning and rearing is still in their infancy especially for grey mullets (*Mugil* sp.) and milkfish (*Chanos chanos*).

The anothers have reviewed the methods of seed production and destitution of both fresh and brackish water cultivable species. They also discuss problems of (*i*) egg, spawn fry and fingerling collection from natural waters and those connected with the spawing of carps under artificial conditions, (*ii*) seed production of carp fish, and other air-breathing fishes, tilapeas elbs (*Anguilla* sp.), and (*iii*) seed production of the brackish water and marine species, viz grey mullets, milkfish and yellow tail (*Seriola quinqueradiala*).

Various issues dealt with in this paper are:

- ☆ Production of seed of Freshwater fishes
    - ☆ Collection of spawn, fry and fingerlings from natural water
    - ☆ Problems in fish breeding in confined waters through controlled natural and seminatural conditions.
    - ☆ Production seeds of catfishes and their air breathing fishes.
    - ☆ Seed production of Channel catfish.
    - ☆ Production of seed of Tilapias
- ☆ Production of seed of brackish water and marine fishes (Grey mullet seed milkfish seed, yellow tails seed.)

The scientists felt that inspite of some advances made in the methods of fish seed production and distribution there remain major constraints on the expansion of fish culture. Supply of pituitary glands for breeding has been inadequate. They think it would be better to find a chemical substitute with a standard potency. The Stage of maturity of fish is also not much known from external morphology on the size of egg. To determine the dosages to be injected.

Methods of inducing natural spawning through hypophysation in Chinese carps have yet to be devised. Methods to bring about proper gonadal development and early maturity or alternatively reduce the fat content should be studied and techniques of raising fish separately. For breeding and for consumption as in poultry production need development. Production of milkfish and prawn seed on a commercial scales is lively to remain a problem until their species can be grown to sexual maturity under controlled conditions. As it is collection of milkfish spanners and keeping them alive for induced breeding itself farms a major problem. Non availability of efficient and cheap feeds for mullet larval one other important problems hampering large scale production.

Such fish culture also aims at improving the production, selective breeding and hybridization for obtaining all male or all female broods, and varieties of seed resistant to fluctuations in the environment need to be developed, continued research is also required on the diseases affecting the egg, level and fry stages of almost all species.

References

Alikunhi, K.H., Improving the breed of Carp. *Bamidgeh*, 8(1): 14-7.

Alikunhi, K.H., Notes on the bionomics, breeding and growth of the murrel, *Ophicephalus*.

Alikunhi, K.H., 1957. Fish culture in India. *Fm Bull*. Delhi (2):144 P.

Alikunhi, K. H., 1960. Practical hints on the breeding of common carp. Publ. cent. *Inl. Fish. Res. Inst*. Brackpore, II P. (mimeo)

Alikunhi, K.H. and H. Chaudhuri, 1959. Preliminary observations on hybridization of the common carp (*Cyprinus carpio*) with Indian carps. *Proc. Indian Sci. Congr.*, 46

Alikunhi, K.H. and K.K. Sukumaran, 1964. Preliminary observations on Chinese carps in India. *Proc. Indian Acad. Sci.*(B), 60(3): 171-88

Alikunhi, K.H., H. Chaudhuri and V. Ramachandran, 1952. Response ot transplantation of fishes in India, which special reference to conditions of existence of carp fry. *J. Asiatic Soc. Beng*. 18(1): 35-53

Alikuhni, K.H., K.K. Sukumaran and S. Parameswaran, 1963a. Induced spawning of the Chinese carps *Ctenopharyngodon idellus* (C. and V) and *Gypophthalmnichthys molitrix* (C. and V) in ponds at Cuttack, India. *Curr. Sci.* 32: 103-106.

Alikuhni, K.H., K.K. Sukumaran and S. Parameswaran, 1963b. Induced spawning of the Chinese grass carp, *Ctenopharyngodon idellus* (C. and V.) and the silver carp. *Hypophthalmichthys molitrix* (C. and V.) in ponds at Cuttack, India. Proc. *Indo-Pacif. Fish Coun.*, 10 (2): 181-204

Alikunhi, K.H., M.A. Vijayalakshamanan and K. H. Ibrahim, Preliminary observations on the spawning of Indian carps, induced by injection of pituitary hormones. *Indian J. Fish* 7:1-19

Alikhunhi K.H. *et al.* Preliminary observations on commercial breeding of Indian carps under comtrolled temperature in the laboratory. *Bull. Cent. Fish. Res. Inst.*,

Atz, J.W. and G.E. Pickford, 1959. The use of pituitary hormones in fish culture. *Endeavour*, 18(71):125-9

Barrackpore, Central Inland Fisheries Research Institute, Rep. cent. *Inl. Fish. Res. Inst., Barrackpore*, 1960-61 (mimeo)

1962. Red. Cent., *Inl. Fish res. Inst. Barrackpore.* 1961-62 (mimeo)

1963. Red.Cent., *Inl. Fish res. Inst. Barrackpore.* 1962-63 (mimeo)

Chaudhuri, H., 1956. Effect of pituitary injections and other factors on reproduction of pond fishes. *Ala Pol. Inst. Bull.* 1956(M.S. Dissertation)

Chaudhuri, H., 1959a. Notes on external characters distinguishing sex of breeders of the common *Indian Carps. Sci. and Cult.* 25(10):258-9

Chaudhuri, H., 1959b. Experiments on hybridization of Indian carps. Proc. *Indian Sci. Congr.* 46

Chaudhuri, H., 1959c. Methods followed in inducing spawing of Indian major carps by pituitary injection. *Publi. Cent. Inl. Fish. Res. Sta.* Barrackpore. 9 P. (mimeo)

Chaudhuri, H., 1960. Experiments on induced spawning of Indian carps with pituitary injections. *Indian. J. Fish.* 7:20-48

Chaudhuri, H., 1961. Spawning and hybridization of Indian carps. *Proc. Pacif. Sci. Congr.* 10 (abstract only)

Chaudhuri, H., 1962. Induced breeding and development of common catfish, *Ompok bimaculatus* (Bloch.) *Proc. Indian Sci. Congr.* 49

Chaudhuri, H., 1963. Induced spawning of Indian carps. *Proc. Nat. Inst. Sci. Indian* (B) 29(4): 478-87

Chaudhuri, H., 1964a. Introduction of exotic species of fish and their effect on the culture of indigenous species of Indian. *Seminar Inland Fish. Dev.*, Uttar Pradesh, Lucknow, PP. 83-92

Chaudhuri, H. and K.H. Alikunhi, 1957. Observations on the spawning in Indian carps by hormone injection. *Curr. Sci.* 26: 381-2

Chaudhuri, H., S.B. Singh and K.K. Sukumaran, 1966. Experiments on large scale production of fish seed of the Chinese grass carp, *Ctenopharyngodon idellus* (C. and V) and the in India. *Proc. Indian Acad. Sci.*(B), 63(2): 80-95

Chaudhuri, H. *et al.*, 1962. Induced breeding of carps during 1960 season at Cuttack. Proc. *Indian Sci. Congr.* 46.

Donaldson, L. R, D. D. Hansler and T. N. Buckridge, Inter-racial hybridization of cutthroat trout salmo clarkia, and its use in fisheries management. Trans. Amer. Fish. Soc. 86:350-60.

Chapter 8

Aquaculture and Rural Communities: A Concept that Struck Prof. Chaudhuri Three Decades Back

A RESUME

During past two decades extensive work has been done to record the symbiotic relationship between aquaculture and rural community interactions and relationships.

Prof. Chaudhuri's scientific work and the application of the result to the benefit of village communities had the basis of his missionary zeal to serve the poor people which has been reflected in the article.

The professor was conscious of the problems of developing countries which have otherwise great potential for aquaculture. He felt the socio-laboratory concept might improve the socio-economic conditions of the rural people.

Prof. Chaudhuri's work detailed hereinafter was written in 1985 when he was at SEAFDEC (Tigbauan). Some portions of his published article entitled. "Developing Rural Communities living at the water's edge, through aquaculture" have been reproduced in this presentation.

1. Background

Man's steady encroachment is rapidly changing the natural environment. The natural environment is being degraded.

Environmental research is now broadened as the environmental scientists have observed the modern trend of industrialization and environmental abuses are leading the world at the edge of crisis. This environmental dilemma is the result of four major factors:

1. Rapid increase of population
2. Environmental pollution
3. Excessive exploitation of natural resources, and
4. The gradual deterioration of a land ethic.

While International Union for Conservation of Nature and Natural Resources (IUCN)-supported natural conservation strategies are of assistance to maintain essential ecological processes and life support systems, it is very often extremely difficult to enforce suitable conservation measures because of growing human pressure, poverty, scarcity of food and other constraints. It is therefore essential to provide suitable alternatives to the destructive use of natural resources. The problem is becoming particularly acute with the aquatic environment.

With the tremendous population explosion, land area is exploited to the maximum for food production and people are forced to utilize natural resources without giving any consideration to conservation measures. Men are rather forgetting the "basic principle of saving some of to-day's resources for tomorrow". The pressure on lands has led men to look towards the ocean for food. It is prophesied that in the future increased dependence on waters for food will alter the nutrition of the millions and the society will shift its reliance from Agriculture to Aquaculture. During a recent (1984) World Conference in Fisheries Management held at FAO, Rome, aquaculture was identified as one of the major strategies to be adopted to accelerate world fisheries production. Our marine catches are getting depleted fast due to over fishing and several other constraints, such as, increasing cost of fuel, dearth of foreign exchange and restrictions due to recently implemented sea law. The vast mangrove areas, which are the nursery grounds for many commercial marine fishes, are being destroyed or reclaimed for other purposes. To supplement production from commercial fishing, the alternative is to develop aquaculture technology in marine, brackish-water as well as in freshwater environments.

Establishment of a social-laboratory (community project) using aquaculture technique as the vehicle for community development help in the conservation measures in water bodies and at the same time will supply protein to the undernourished population and also provide additional income to the rural community living at the water's edge. In this project the rural community is focused as the subject and object of development, and aquaculture as the means for such development. The programs will be mainly taken up in areas which are potentially viable for aquaculture development.

2. Social-Laboratory

A social-laboratory is a pilot community project on rural development. At the community level the citizenry will have to develop a deep concern and take a keener interest in programs on rural environmental problems. A novel approach to environmental problems is to get those affected in the village community involved in an action oriented program like aquaculture in the waters adjacent to the area where the community lives. These communities at the grass-root level can make significant contributions to mend ecosystems.

Any portion of our natural environment that man utilizes to promote his welfare can be identified as a natural resource. Coastal areas, mangroves, lagoons, estuaries, rivers, streams, lakes, inundated low lying areas, swamps, canals and creeks are the kind of natural resources considered in this project. Harmony between man's resource requirements in general, and the dire need of the millions below the poverty level in the country side in particular, and the resource base depends upon sound policies of better resource utilization that will serve both environmental conservation and a livelihood generation to the most impoverished sector of the society.

3. Objectives

The general objectives of a Social-laboratory project are:

1. To develop self-reliant and self-sustaining rural communities through the active involvement of and meaningful participation by the communities in their own development.

2. To develop strategies to facilitate transfer of technology from researchers to end-users.

3. To bring satisfactory change and growth to rural life.

The more specified objectives are:

1. To help the community residing near a water body to become efficient and effectively participate in utilizing aquaculture techniques for increasing aquaculture production and income of the cooperating families.

2. To serve as a demonstration are and also a training centre for the community to develop improved aquaculture technology.

3. To demonstrate the feasibility of aquaculture as a viable socio-economic program for the improvement of the quality of life of the rural community through self-help endeavors.

4. Aquaculture Potential

Fish is one of the main basic items of food of the majority of the people of the Third World and is considered second only to rice in importance. It is a source of low-cost protein and contains the essential amino-acids in desirable concentrations in addition to vitamins, minerals and fats. The present average per capita consumption of fish is only about 8.4 kg although aquaculture potential is very high in these countries. Increasing production of fish in rural areas where more than 75 per cent of the population live will be a substantial contribution towards human nutrition as well as income generation.

Large areas of inland and coastal waters in the world, especially in the developing countries of Asia, Africa and Latin America are lying practically untapped and are not properly utilized for fish production. Vast majority of these water bodies are either over-exploited or rendered unproductive due to environmental pollutions caused by industrial, agricultural and domestic wastes. The extensive areas of coastal waters including protected covers, bays, tidal flats, lagoons, mangrove swamps and estuaries offer great potential for culturing finfish, shellfish and seaweeds by the small-scale fishermen communities residing near the coast lines to take up aquaculture activities to supplement their income. Inland areas abound with numerous small and large-size ponds, natural pools, man-made reservoirs and lakes, rivers and streams, canals and creeks which could also be utilized for aquaculture by the rural communities residing in the neighborhood. In addition there still exist extensive potential areas of swamp lands, low lying inundated flood plains and shallow impoundments mainly in Bangladesh, India, Burma, Thailand, Vietnam, Indonesia and in the African Plains. In Southeast Asia alone there are some 19 million ha of inland waters available for aquaculture purposes. In China, there are 20 million ha of freshwater areas of which 7 m ha are suitable for fish farming. Of this, only 1.3 m ha are under cultivation. In India, it is reported that there are about 1.6 m ha of freshwater ponds, 2.6 m ha of reservoirs and lakes and the estuaries, brackish-water and backwaters constitute 2.6 m ha. These vast potential areas offer great scope for the development of aquaculture. In addition, there exist about 80 m ha of irrigated rice fields which if utilized for rice-fish culture will give an estimated production of 2.4 m metric tons of fish/yr calculated at a very conservative yield of a meagre 30 kg/ha/yr of fish. Harnessing even a small percentage of these potential areas by developing appropriate aquaculture techniques will help improving the nutritional level of the undernourished rural communities, their poor lot further aggravated by an expanding population.

5.   Aquaculture Technology

The aquaculture technology to be implemented in the community project should not be a sophisticated one; on the other hand the type of technology which needs less-inputs is preferable. The technology should suit the socio-economic condition of the rural communities and also should be appropriate for the aquatic environment. In other words the type of aquaculture technology to be used may vary depending on the type of available aquatic environment. The following aquaculture technologies are recommended for the project depending on the suitability of the water bodies.

Pen Culture

One of the most recent developments in aquaculture is pen culture or culture of commodities inside net enclosures. This has attracted worldwide attention in recent years and is suitable for shallow coastal areas, lagoons, fringes of reservoirs and lakes and also in the shallow inundated impoundments. The productivity of pens is claimed to be as high as 10-20 times more than pond culture. The shallow waters of a pen is generally very rich in natural food. A pen enclosure eliminates the predators and unwanted weed fishes, thereby giving protection to cultivated species from predation and provide more natural food for them by removal of the weed fishes, their competitors for food. A pen can be suitably stocked with the desirable varieties of fishes and the number of fish stocked can be regulated and suitable management measures undertaken.

Laguna de Bay, near Manila in the Philippines, is a good example of successful pen culture. The lake is 90,000 ha in area having an average depth of 2.8 m. Although the water is quite rich, the catches were getting depleted due to mainly over fishing. Pen culture was initiated in 1973 when 4800 ha of pens were constructed. Milkfish was cultured in the fish pens and the production was 4 ton/ha/yr as compared to 245 kg/ha/yr for the open waters. The yield from 4800 ha of fish pens was equivalent to the production of the rest of 85000 ha of open water. The high production in pens encouraged new fish farmers and by 1974 the area under fish pens was increased to 7,000 ha.

Cage Culture

Cage culture of fish has become very popular in recent years and is practiced in many countries for increasing production of fish. This concept of raising fish in cages, however, is not new. Cage culture has been traditionally used for over a century in Kampuchea and later on expanded to the other Southeast Asian countries like Vietnam, Thailand and Indonesia. Subsequently, the application of this technique has been spread to Japan, China, USA and Europe. Cage culture has several advantages. They are:

1. It permits the more intensive use of existing water areas, thereby reducing pressure on land resources.

2. It takes advantage of good water quality of large bodies of inland waters and open seas with adequate circulation of water thereby ensuring adequate supply of oxygen and elimination of metabolites.

3. In deep lakes and reservoirs where thermal stratifications occur and bottom water becomes anaerobic, the upper layers rich in natural food and with higher oxygen content can be utilized for cage culture.

4. It increases survival rate of fish by eliminating predators.

5. It cuts down expenses and hazard of harvesting of fish.

6. The management is easier and the health as well as growth of fish can be checked occasionally.

7. Cages could be removed easily if the culture area gets polluted or threatened by natural hazards.

8. There is optimum utilization of artificial feed for growth and it reduces the length of rearing period.

The main constraint of cage culture is the initial expenditure for construction; but the cost can be cut short by using bamboos and other cheap local materials.

Depending on the suitability of the aquatic environment the cages for culture may be of the following types:

1. Fixed cages resting on the bottom and reaching the water surface;

2. Submerged cages either resting on the bottom or floating in mid-water;

3. Cages floating at the water surface.

Selection of proper site for cage culture is very important. The location should be protected from strong wind or current but there should be free flow of water to maintain optimum dissolved oxygen content. The rate of stocking in cages is very high. The production from cages is manifold as compared to ponds. Depending on the rate of flow of water in running water cage culture with heavy stocking and intensive feeding of balanced complete feeds and constant elimination of nitrogenous waste products, the production could be even over 4000 tons/ha/yr which appears unbelievable. In waters rich in natural food, very high production can be obtained even without any supplementary feeding.

In Singapore, cage culture of bighead carps in productive freshwater reservoirs gave an estimated production of 296 metric tons/ha/yr. Japan raises 6 millions tons of yellowtail by cage culture in coastal waters for sea-ranching purposes. The bird sanctuaries or the lakes visited by migratory birds (aquatic) are very rich in natural fish food organisms due to fertilization by their fecal matters. These waters are ideal for cage culture where feeding is not necessary.

Integrated Fish Farming

Integration of fish farming with livestock and crops productions is a traditional practice in China and several other Asian countries. For centuries small-scale farmers of these regions have sustained themselves by combining fish culture with various crop farming and raising domestic animals such as, pigs, cattle, ducks and poultry. In this system there is mutual benefits and wastes of one commodity is utilized by the other and practically nothing is wasted. This concept of an all-round development of aquaculture, agriculture and animal husbandry has met with tremendous success especially in the communes of modern China and can play a major role in the rural development programs of other developing countries of the world. However, this system has to be modified for application in the developing countries where emphasis should be in the development of small-scale integrated fish farming suiting to the socio-economic conditions of the rural people for integration into the rural economy. Small-scale labor intensive form of commercial aquaculture usually dominates the aquaculture industry of many developing countries.

The rural communities who will be involved in this project could take advantage of this integrated farming systems for meeting self-sufficiency in food and generating additional income and employment. In Lao PDR an FAO/UNDP aquaculture project has demonstrated that an average production of about 3.0 metric tons/ha/yr of fish can be produced by integrated fish farming with livestock and crop productions with practically no inputs in terms of fertilizer and feed for raising fish.

Mussel, Oyster and Seaweed Farming in Coastal Waters

Potential areas for farming mussels and oysters are the protected coves, bays and estuaries. Farming of mussels and oysters is quite simple and can be taken up by the fishermen communities residing near these areas. The cost of investment is very little and large quantities can be grown with minimum inputs. Another advantage of oyster and mussel culture is that their farming does not require skilled manpower. Selection of site however, is important. Seaweed culture also can be taken up in suitable shallow coastal areas adequately protected from storms. Like oyster and mussel, seaweed culture also requires very little capital investment.

6.   Suitable Fish for Culture

While selecting fish species suitable for aquaculture in rural areas preference is given to those which are quick growing, efficient convertor of natural food could be raised at a relatively low cost and are well accepted by the rural communities. The main emphasis should be given on the low-priced fishes which can convert animal wastes to fish flesh and give higher production for the benefit of luxury commodities with longer food chain and raised on artificial feeds with high protein-contents. The common carp and the Asiatic carps are ideal for freshwaters. Also, selected quick growing species of tilapia, omnivorous in food habits and can be raised easily giving high yields, are preferred. The common carp (several strains) is a domesticated species cultured almost all over the world and especially used in cage culture giving extremely high production. Also, various species of fast growing Asiatic carps, such as, bighead, silver carp and grass carp (Chinese carps), Calta, rohu and mrigal (Indian carps) and tawes (Indonesia carps) utilize mostly the primary production planktonic food, detritus and vegetative matters and are cultured in Asia and many other countries. Africa is rich in having several indigenous species of the tilapia group most of which are popular with the natives. Of these the quick growing Nile tilapia is cultured all over the world and is an ideal species for cage culture and in rice-fish culture. For cage culture, the requisite quality of a fish to be selected is that it should be hardy, should be able to thrive in low oxygen content, grow quickly and easy availability of its seed in sufficient numbers. Nile tilapia fulfills all these essential requirements.

For culture in coastal lagoons and brackishwaters, the euryhaline species of fishes, such as, milkfish and mullets could be selected. The rabbitfish (siganidas) can be cultured in coastal waters, tidal flats and mangrove areas. Nile tilapia can also be cultured in brackishwater areas. Other species which are cultured in purely marine environment are the groupers, snappers, sea bass and yellowtails which are high-priced species.

In conclusion, it may be summed up that the developing countries have great potential for aquaculture and also possess efficient species for culture. At present the vast majority of these areas are either unutilized or underutilized and can be developed with little effort. The Social- Laboratory concept as described in the foregoing can be used to improve the socio-economic conditions of the rural communities living at the water's edge through the vehicle of aquaculture by providing them better nutrition, additional income and better employment opportunities.

Chapter 9

More Information on Carps

A RESUME

Prof. Chaudhuri's elaborate work on Carp has been mentioned in the earlier publication entitled, "Aquaculture Beyond 2000: New Horizon". This chapter records a few more observations made by him supplemented by observations made in "Fishing Chimes". K. R. Prasad (Fishing Chimes, April, 2007) observes, "The impact this technology induced breeding developed by Dr. Hiralal Chaudhuri was historic. It spread instantaneously over all the states of India, studding the nation at all locations convenient for the purpose with a number of major carp hatcheries. These hatcheries generated a well-sustained wave of major carp seed production. As a result, there has been a beneficial effect which was one of discouraging major carp spawns collections from rivers. This contributed to the progressive restoration of major carp populations in them."

Prof. Chaudhuri's elaborate work on Carp has been mentioned in the earlier publication entitled, "Aquaculture Beyond 2000: New Horizon". This chapter records a few more observations made by him supplemented by observations made in "Fishing Chimes". K. R. Prasad (Fishing Chimes, April, 2007) observes, "The impact this technology induced breeding developed by Dr. Hiralal Chaudhuri was historic. It spread instantaneously over all the states of India, studding the nation at all locations convenient for the purpose with a number of major carp hatcheries. These hatcheries generated a well-sustained wave of major carp seed production. As a result, there has been a beneficial effect which was one of discouraging major carp spawns collections from rivers. This contributed to the progressive restoration of major carp populations in them."

More Research on Carp and Benefit Therefrom

This chapter records observations made by Dr. K. R. Prasad, Dr. S.K. Mukhopadhyay, Dr. C. P. Singh *et al.* and Dr. Dixitulu. These observations link Prof. Chaudhuri with various findings based on their analysis. Besides, a few works of Prof. Chaudhuri have also been included by some experts in this chapter which are as follows:

1. Major Carp: Export Dimension by Dr. Prasad.

2. Seed Production through hypophysation by Dr. S. K. Mukhopadhyay

3. Achievement of CIFRI by Dr. M. Sinha.

4. Induced Breeding of *Labeo rohita* by Dr. C. P. Singh *et al.*

5. Contribution to Fresh water Aquaculture by Chief Editor, Dixitulu

6. External characters distinguishing sex of breeders of the common Indian Carps

7. Role of sex specificity of carp pituitary gland in induced spawning of carps by Dr. K. B. Ibrahim and Prof. Chaudhuri

8. Carp hatcheries development by A. K. Reddy in Fishing Chimes (Jan-Feb 2000)

In the book *"Aquaculture Beyond 2000: New Horizon"*, the following chapters have been included which are not being repeated in this treatise:

☆ Induced breeding of carp

☆ Pituitary Gonadotrophin in Indian major carp

☆ Induced breeding and hatchery management in carps

☆ Induced spawning of Indian carps

☆ Selective breeding and hybridization of carp.

1. Carp and Export Dimensions

K. R. Prasad (in *Fishing Chimes* – April 2007) says, "Major Carps are India's pride. In 1950s, this pride acquired added glory with the development of the technology for the induced breeding of these carps. The impact of this technology, developed by Dr. Hiralal Chaudhuri, was historic. It spread instantaneously over all the States of India, studding the nation at all locations convenient for the purpose with a number of major carp hatcheries. These hatcheries generated a well sustained wave of major carp seed production. As a result, there has been a beneficial effect which was one of discouraging major carp spawn collections to the progressive restoration of major carp poulations in them".

He further says. "The main outcome of the spread of the technology of induced breeding, particularly in relation to the major carps, has been the sprouting of their seed for supplies to farmers. What had happened in practical terms was that, with the availability of copious supplies of seed, major carp production increased. Due to this, there had been a spurt in farm production of major carps".

It has to be admitted that, the scenario in respect of major carps exports has been bleak. There is however a ray of hope now. This arises from a feature that has appeared in the U.K. based journal *Seafood International*, October, 2006 issue, published with the caption 'Supplies and Markets' at page 19.

Contrary to the general impression, carp is the most popular freshwater fish after trout in Europe, according to *Seafood International*, October, 2006 issue. Hungary, the Czech Republic, Poland and Germany are the main consumers. The journal says that the Central European Countries are the most promising markets for the fish. Saying that in some areas like the Mississippi delta, carps have become a menace, it was observed in the feature that the solution to get over the problem would be to develop the market for carps so that the increased catches can be disposed of economically. This approach applies equally well to the present problem of increasing catches of major carps faced in India. Pointing out that the negative image westerners have about carps (although they do have a market but with limitations), the journal suggests that carps be called as 'silver cod' so as to generated greater market interest.

Considering the foregoing points. it is desirable the measures are initiated for exporting Indian major carps to countries of Central Europe in the first instance. The measures to be taken can include (*a*) popularising Indian major carp with appropriately designing names that can capture into of the people of that part of Europe. (*b*) sending a team of specialist in Central European countries to discuss the subject of export of Indian major carps with people concerned in those countries to prepare ground for sending consignments for testing major reaction and price aspects; and sending trial consignments to identified agencies in the countries for marketing and further follow-up action.

Establishment of a market of Indian major carps in Central Europe to start with, can well prove to be a major step in the direction of helping our fish farmers upgrading their earnings thereby their economic status.

(Source: Fishing Chimes, April, 2007)

2. Seed Production through Hypophysation

The Indian major carps (*Catla catla, Labeo rohita, Cirrhinus mrigala*) and Chinese carps (*Hypophthalmichthys molirix, Ctenophya-rungodon idella*) normally breed in flooded rivers, during monsoon months. Until the technique of induced breeding came into vogue, the only source of carp seeds for aquaculture in India was rivers.

Induced Breeding Technique

The commonly adopted technique of induced spawning of carps is hypophysation *i.e.*, by injecting the carps with carp pituitary (hypophysis) gland extract. Induced spawning in fish, employing hypophysation as a tool was first developed by the Brazilians (Von Ihering, 1937). Khan (1938) first attempted induced breeding of Indian major carps by using mammalian pituitary. He succeeded in producing ovulation in the fish. Chaudhuri and Alikunhi (1957) first succeeded in induced spawning of Indian major carps – *Labeo rohita* and *Cirrhinus mrigala* by injecting fish pituitary hormone. Chinese carps were successfully induced to breed in 1962 by following the same technique (Alikunhi *et al.*, 1963). Since then the technique has received a wide popularity and is now a part and parcel of aquaculture. [Mukhopaddhyay; ICAR in Fishing Chimes, May, 2000].

The following steps have be to be taken to obtain pure carp spawn through hypophysation as observed by Mukhopadhyay:

How to Produce 10 Million Carp Seed During One Monsoon Season

A trained farmer can handle four sets of brooders per day. He can handle about 60-65 sets during the monsoon seasons (15 to 16 cool rainy days). Two such farmers can handle 120 sets for production of 10 million carp spawn, comprising 1.5 million catla, 2.5 million mrigala, 2.5 million silver carp, 1.5 million grass carp during a single monsoon season utilizing 1,616 kg brood stock (Tripathi, 1981).

Genetic Upgradation of Stocks

Unlike developed countries, there are no recognised strains of our cultured species. Long term genetic gain can only be achieved, if we work with strains with known genomic value, instead of using unknown stocks, which have undergone inbreeding. The application of genetics for strain identification has been taken up in India somewhat later than in other countries. Because of this we are somewhat lagging behind, but we will be able to overcome this bottleneck in the near future through concerted effort of all the scientists in the country in the field of fish genetics (Das and Ponniah, 1996).

(Source: Fishing Chimes, May, 2000)

3. Achievement of CIFRI in Last 50 Years

M. Sinha, Director, observes, "The Central Inland Fisheries Research Institute has left an indelible mark on the inland fisheries scene during the last five decades by developing many epoch-making technologies. They cover all facets of inland fisheries such as fresh and brackishwater aquaculture, mixed farming of fish and prawn aquaculture through waste recycling and integrated farming systems involving fish poultry, duck and pig farming, and rice-cum-fish culture. Similar R&D efforts have been made in the open water system, leading to higher productivity." The major achievements of the Institute during the last five decades are briefly described below M. Sinha:

Aquaculture–Carp Seed Production Technology

One of the factors that retarded growth of aquaculture as a profitable vocation was the inadequate production and supply of quality fish seed. Realising this need, CIFRI directed its research attention towards developing techniques for the breeding and seed rearing of Indian and exotic carps during the 1960s. By standardising the techniques for induced breeding of carps through administration of pituitary gland extract and seed production through various phases of larval rearing the Institute had revolutionised the freshwater aquaculture system in the country. This technique was refined further to enable fish farmers to preserve pituitary gland extract in ampoules for use in remote villages and for industrial production of fish seed by enterpreneurs. The ampouling technique has paved the way for setting up *Pituitary Banks* a prerequisite for small scale as well as industrialised freshwater aquaculture. By the 1980s, use of various substitutes for pituitary such as HCG and synthetic analogue LHRH had been standardised. Another major development in this area was the development techniques for off-season breeding. Through techniques for off-season breeding. Through improved methods, it has been made possible to breed Indian carps as early as February and as late as October.

Carp Culture Technology

The traditional fish culture practised in the eastern region of the country was not scientifically sound, resulting in low yields. CIFRI has in its initial experiments used combinations of Indian major carps (catla, rohu and mrigal) for stocking resulting in production ranging from 1,500-3,000 kg/ha/yr. Later, the species spectrum was broadened to include some exotic species such as grass carp, silver carp and common to utilise the unused and unshared niches in the pond ecosystem. Based on the echo-niche approach, the Institute developed a mixed carp culture system which had later become popular as the Composite Fish Culture. This was a major landmark in the history of aquaculture in the country as the yields went up to 10t/ha/yr, against 600 kg/ha/yr under the traditional system. Later this technology was demonstrated nation-wide in the farm conditions of various geo-climatic regions of the country under All Indian Coordinated Project. All these culminated in the *blue revolution*, witnessed by the country by the end of 1970s. This technology continues to be the main framework on which the country's major freshwater aquaculture development plans like Fish Farmers' Development Agencies (FFDAS) operate even today.

Integrated Farming System

Productive utilisation of farm wastes as inputs in agriculture and animal rearing is recognised as a means of efficient land use under low cost farming systems. Towards this goal, system of integrating composite fish culture and livestock farming had been evolved for wider adoption by marginal farmers and tribals. In view of high compatibility of the ducks with the stocked fishes, the raising of ducks in and around a fish pond fits very well with the concept of tegrated approach in resource utilisation. The ducks do not compete for food with the stocked fishes, as they live on the larvae of the aquatic insects, tadpoles, molluscs and a few varieties of the aquatic weeds. At the same time, duck droppings are directly utilised by fishes as food and they also serve as excellent fertiliser in ponds. Duck-cum-fish culture system ensures a yield of 2230-3941 kg fish, 14,000-32,256 eggs and 400-750 kg duck meat from a one ha. farm.

The technique of fish-cum-farming developed by the Institute has opened up new horizons of producing high quality animal protein at a low cost. This technology has proved to be an efficient means of waste utilisation with complete saving on fertilisers and supplementary feed for fish production. The waste material i.e., pig dung acts as a substitute for fish feed and pond fertilisers. The system ensures fish production of 6644-7306 kg/ha/yr, besides 5610-10,950 kg of meat.

Growing paddy along with fish and prawns has been an age old practice in many parts of the country, but the yields obtained from such systems were not very encouraging. Many systems of paddy-cum-fish culture were developed at CIFRI to suit various situations prevailing in more than 2.3 million ha of lowlying areas of the country where rice fields retain sufficient water during the monsoon. Culture of fish along with deep water paddy yields better returns to the agriculturists, since, in addition to paddy some fish also are produced. Per hectare yield obtained under paddy-cum-fish culture is 700 kg/10 months kg fish and 1200 kg/ha (kharif) 4,300 kg/ha (rabi) paddy.

Fish Culture Sewage-fed Ponds

At a time when the disposal of sewage was nagging problem in the cities and industrial townships of India, CIFRI had shown a way to recycle these wastes through aquaculture, the Institute had developed a production technology using treated sewage as a source of nutrients. Studies had shown that a gross production of 9,350 kg/ha/yr could achieved using *Tilapia mossambica*. The technology had paved the way for recycling domestic wastes through fish production and helping the efforts to abet the environmental degradation through sewage wastes.

Aquatic Weed Control

Another landmark achievement of the Institute is the development of technology to control aquatic weeds in aquaculture systems. Different methods for manual, mechanical, chemical and biological control had been standardised for use in various kinds of water bodies ranging from small ponds to lakes.

Low Input Technology for Air-breathing Catfishes

Low input culture technology for culture of air breathing catfishes *viz. Clarias batrachus* (magur) and *Heteropneustes fossilis* (singhi) was successfully demonstrated in the farmers' shallow seasonal ponds in West Bengal. In regard to the positive response of the air breathing catfishes *Clarias batrachus* and *Heteropneustex fossilix* to assimilate non-protein nitrogen, the ultimate evidence of the inherent biochemical pathways leading to the utilisation of ammonia nitrogen from urea had been ascertained through the mass spectrometric studies using 15N-Urea in the diet. This research finding has far reaching applicability as feed combinations with non-protein nitrogen can partially replace the animal protein component in the diet helping considerable cost reduction in cat fish culture operations.

Brackishwater Aquaculture

Nursery and rearing practices ensuring high survival of the young fish and prawn have been developed for adoption by fish farmers. It was made possible to obtain a survival of 75 per cent at a stocking density of 10,000 post-larvae of tiger shrimp/m^3 in plastic pools. A survival of 75 per cent and 41 per cent respectively of the mullets *Liza parsia* and *L. tade* at 1,00,000 and 90,000/ha and 90 per cent of the perch *Lates calcarifer* at 1,00,000/ha was also ensured in the nurseries. Monoculture of shrimp/fish and polyculture of fish and shrimps have been developed.

(Source: M. Sinha, Director, CIFRI; Fishing Chimes, April, 1997)

4. Induced Breeding of *Labeo rohita* Using Ovaprium and Pituitary Gland Extract

The following observation of Singh *et al* signifies that development of aquaculture gained a momentum since 1957 the year induced breeding of Carp got a success through the hypophysation by Prof. H. Chaudhuri.

Availability of quality fish seed in adequate numbers and their stocking by farmers in their ponds at appropriate time substantially enhances fish production in their ponds. Prior to 1957, when the major source of seed procurement was from the riverine resources, carp culture was confined to small scale operations, owing to the poor quality of fish seed. The seed collected from the natural resources often got mixed with the seed of undesirable species, resulting in poor culture fish production.

With the introduction of induced breeding technique of Indian Major carps by Chaudhuri and Alikunhi (1957) and exotic carps by Alikunhi *et al* (1963) through hypophysation technique has been standardised and refined with the use of several alternative spawning agents to pituitary gland such as HCG, LR-RH, Ovaprim, Ovatide etc., with varying degrees of success. The superiority of Ovaprim over pituitary extract inducing breeding of major carps has been reported by several workers (Nandeesha *et al.*, 1990; Fancis *et al.* 1991 etc.). However, the studies conducted so far have been limited to examine the effect of different hormones on the performance of spawning and hatching only. The influence of these hormones on further growth stages of fish has not been evaluated.

(Source: Fishing Chimes, July, 2002)

5.   Carp and Freshwater Aquaculture

Freshwater Aquaculture

Prof. Hiralal Chaudhuri was a pivotal figure in CIFA (Central Institute of Freshwater Aquaculture). Fishing chimes has summarised CIFA's (Central Institute of Freshwater Aquaculture) contribution to aquaculture much of which was primiarily due to research success of Professor Chaudhuri as follows:

The CIFA (Central Institute of Freshwater Aquaculture) has, as per the mandate, been engaged in research in freshwater aquaculture, in the disciplines of Production Technology, Soil-Water Environment, Fish Genetics, Fish Nutrition, Fish Physiology, Fish Pathology, Aquaculture Engineering, Aquaculture Economics and Statistics, and Aquaculture Extension as thrust areas, along with training of manpower and extension. The Institute, being a Regional

Lead Centre for Carp Farming under the Network of Aquaculture in Asia-Pacific (NACA), has hosted three UNPD/FAO programmes. The Project on Centre of Advanced Studies (CAS) in Freshwater Aquaculure, in collaboration with Orissa University of Agriculture and Technology, Bhubaneswar, provides for award of M.F.Sc. and Ph. D degrees. The institute offers training programmes in different aspects of freshwater aquaculture and has conducted seminars/ workshops at frequent intervals. Along with the Extension Section, the Institute also has the provision for transferring the different technologies through KVK/TTC and Kausalyaganga. Thus the Institute is carrying out different aspect of research, training and extension in freshwater aquaculture, fulfilling the objectives.

Since the establishment at Cuttack in 1949, the Division had made epoch-making contributions to the development of fish culture especially in carp breeding, seed raising and table-fish production. In fact two major technologies of induced breeding and composite carp culture that laid the foundations of the modern Indian aquaculture were developed at this Division. This has led to a virtual change in the scenario of fish seed production in the country from total dependence on riverine seed collection to production of over 15,000 million nos., of farm-bred fry at present. It is remarkable to note that one-third of the total fish production in the country today comes from freshwater aquaculture alone, which is the result of catalysis by the technologies developed at the Institute. The Division has several credits in terms of development of specific packages of freshwater aquaculture technologies like sewage-fed carp culture, peninsular tank fisheries, integrated fish farming with agriculture and livestock etc. As a reorganised Institute, CIFA has made several outstanding contributions such as multiple breeding of carps, gamete cryopreseration, intensive carp culture with production rates of 10 and 15 t/ha/yr, and breeding and hatchery management of catfish, *Clarias batrachus* and freshwater prawns, *Macrobrachium rosenbergii* and *M. malcomlmsonii*, production of cultured freshwater pearls through nuclei implantation in freshwater mussels, aquatic biofertilisation with *Azolla* and utilisation of biogas slurry and organic farming practice, production of sterile triploid grass carp, formulation of diets for fish and prawn species (CIFACA–as commercial carp diet), formulation of CIFAX as a control measure of epizootic ulcerative syndrome, packages of practices for sewage fed fish culture and cage culture, breeding and rearing of commercially important frog species etc., which are a few to name.

The Institute has generated several technologies that have paved way for the entry of new enterprises into the line of freshwater aquaculture and much of the new developments in freshwater aquaculture sector have been the result of catalysis ushered in by the achievements of the Institute. Having achieved high carp production levels of 15 tonnes/ha/yr, the Institute is presently formulating a national plan for freshwater aquaculture development on a district basis for raising average national productivity to 5 tonnes/ha/yr. Intensification and diversification of aquaculture systems with mixed culture of different species, organic recycling for sustained productivity, integrated aquaculture management, etc., are being promoted. Greater emphasis is being laid on technology transfer through intensive training programmes, implementation of Operational Research Project, on-farm demonstrations, consultancies and publications. Freshwater aquaculture holds the key for increasing fish production in the year to come CIFA is playing its role in increasing productivity on a sustained basis not only within the country but in the region.

(Source: Dixitulu in Fishing Chimes, July 2002)

6. Notes on External Characters Distinguishing Sex of Breeders of the Common Indian Carps

The success of the recent method inducing breeding of Indian carps by injection of pituitary gland hormones, depends to a large extent of the correct identification of the sex and stage of maturity of the breeders to be injected. The Indian carps, all from natural wild stock, are very active fishes which jump widely and often injure themselves when netted or kept in a confined space. When used for breeding purposes such large active fishes have to be very carefully handled to avoid injuries. To distinguish the sex such distinguishing characters have to be found which could be easily and quickly observed and which will not necessitate keeping the fish out of water for any appreciable length of time.

In bony fishes, well-marked structural differences distinguishing the two sexes are usually not very common except in a few Cyprinodonts in which the intromittent organs of the males easily mark them out. Secondary sexual characters are, however, seen in the majority of oviparous Telecosts and these may be present in the adults throughout the year some and only during the breeding season in others.

In general, in the great majority of the bony fishes, the females are larger than the males, although in some fishes especially among those having parental care (*e.g.* Murrels, Cichlids, etc.) the males are found to be larger than the females. In cyprinids, the secondary sexual differences during the spawning season are usually confined to the bright colouration of the males, appearance of horny tubercles on the snout, differences in size, shape and length of the fins and roughness of the operculum and pectoral fins. In addition to these the fully mature males could be identified during the breeding season by slightly pressing the abdomen when milk-like fluid oozes out of the vent. In many of the Cyprinids, the gravid females are identified by the distended, bulging, roundish abdomen and the swollen reddish vent.

Russian fish culturists distinguish females of the common carp, *Cyprinus carpio* from their distended abdomen and swollen reddish vent. When such females are laid on their back on a table a narrow crease appears along the length of the stomach due to the sinking of the egg sac into the abdominal cavity. This feature is absent in the males which can also be identified by the perceptible roughness of the back which can be felt when the hand is run along from the tail-end to the head. This roughness is especially prominent on the head and gill covers.

In goldfish, *Carassius auratus*, the mature males are distinguished from the females by the roughness of the gill-covers and whitish prominent denticulations on the pectoral spines. In *Barbus dubius* the male is distinguished by the presence of nasal pores on the snout and the female has a longer anal fin which, when distended, touches the caudal fin. In *Labeo dero* and *Labeo boggut*, the adult male develops prominent horny tubercles on the snout as the spawning seasons approaches and the transverse groove on the snout becomes comparatively much deeper than in the female. The mature males of *Labeo dero* are also distinguished from the females by the greatly elongated anterior rays of the dorsal fin.

Major carps, which constitute the important food fishes of India, do not have any prominent structural sexual differences from which sexes could be easily distinguished. According to Mookerjee, in

L. rohita males the pectoral fin is either longer or equal to the anal fin whereas in the femals the pectoral fin is shorter than the anal fin. In *C. mrigala* on the other hand, the pectoral fin is longer than the anal in males and equal to the anal in females. Study of the sexes based on the above characters is not practicable in the field, as it is very difficult to take the comparative measurement of fins in large size live breeders.

During the 1957 carp breeding season attempts were made to find out some field method identifying the sexes of the major and minor Indian carps. It was found that the pectoral fin in mature males of the majority of the species is slightly stouter and when extended backwards and towards the dorsal side of the body meets the lateral line scales at a place more posterior to that in the mature females. Thus in the female if the tip of the pectoral fin reaches the x-th lateral line scale counted from the beginning, it extends in males up to the $(x + 1)$th to $(x + 3)$th scale.

Identification of the males and females of some of the important major carps and minor carps during the breeding season, based on the above character, is given in the following table:

Species	Sex	No. of Breeders examined	Pectoral fin reaching the lateral line scale	
1. *Labeo rohita*	Male	55	8th or 9th 1.1.	scale
-do-	Female	80	6th or 7th	-do-
2. *Cirrhina mrigala*	Male	23	10th or 11th	-do-
-do-	Female	32	8th or 9th	-do-
3. *Labeo bata*	Male	39	8th	-do-
-do-	Female	42	7th	-do-
4. *Cirrhina reba*	Male	67	8th or 9th	-do-
-do-	Female	68	6th or 7th	-do-
5. *Catla catla*	Male	3	8th or 9th	-do-
-do-	Female	3	6th or 7th	-do-
6. *Labeo gonius*	Male	15	18th or 19th	-do-
-do-	Female	12	16th or 17th	-do-
7. *Labeo calbasu*	Male	13	10th or 11th	-do-
-do-	Female	14	8th or 9th	-do-

While an appreciable number of mature specimens of the first four species was examined and the character was found to be true in more than 95 per cent of the cases, the number of specimens examined of the last three species is inadequate for making any generalization.

During the 1958 carp breeding season, a much easier method of identification of sexes in mature carps was found out. The pectoral fin in mature males has a very rough dorsal surface (*i.e.* surface close to the body) which in the case of the females is very smooth. The difference can easily be foundout just by feeling the dorsal surface of the pectoral fin by the finger tips. This distinguishing character is present during the breeding season in all mature specimens of *Catla catla, Labeo rohita, Cirrhina mrigala, Labeo calbasu, Labeo gonius, Labeo bata, Labeo pangusia* and *Cirrhina reba*. It is believed that this feature might be present also in other Indian carps of similar breeding habits. The rough surface of the pectoral fin could be of help in forming a grip of the ventral side of the female at the time of the courtship and sex play.

When sexing of breeders is to be done in the field of purposes of inducing breeding, the brood fishes hauled, are taken one by one in deep bag like hand nets (made of fry-net bags stitched to a bamboo or crane frame with handle) and the sex determined quickly by inserting the hand in the net bag and feeling the dorsal surface of the pectoral fin with the fingers. The males and females are then kept in separate ponds or enclosures. Taking the fish in net bags prevents it from struggling too much and injuring itself. When injected breeders are to be taken out for subsequent injection also, sexing can be easily done by the above method.

(Source: H. Chaudhuri, Science and Culture, Oct, 1959)

Some Selected References

Chaudhuri, H. and Alikunhi, K.H., 1957. *Curr. Sci.*, 26: 381.

Hora, S.L. and Misra, K.S., 1936. *Rec., Ind. Mus.*, 38: 341.

Misra, K.S., 1950. *F. Zool. Soc. India*, 2: 91.

Mookerjee, H.K., 1938. *Ind F. Vet. Sci and Anim. Husb.*, 8.

Norman, J.R., 1948. History of Fishes, p. 299 (A. A. Wyn, Inc., New York).

Schoverchov, F.N., 1953. *Prudovoje Rybowdstyo*, (Compiled by M. Alkunhis, *Bamigdeh*, 7, 6, 1955).

7. Studies on the Role of Sex-Specificity of Carp Pituitary Glands in Induced Spawning of Indian Carps

In view of the differences expressed by several workers on the potency of pituitary glands collected from male and female fish, experiments were taken up to find out whether segregation of glands sexwise is essential for induced breeding work.

Glands were collected separately from males and females of *Labeo rohita*, *Cirrhinus mrigala*, *Cyprinus carpio*, and *Hypophthalmichthys molitrix* and preserved in absolute alcohal. Pituitary gland extract was prepared in distilled water and glycerine in bulk under identical conditions. During the fish breeding seasons of 1966 and 1967, 124 sets of *L. rohita* were injected with sex-segragated gland extracts. Equal number of sets were injected on the same day under identical conditions, separately with male and female gland extracts, and the results obtained compared.

Out of 74 sets injected with rohu gland, 75.7 per cent positive result was obtained with female gland extract and 67.5 per cent with male gland extract. In case of mrigal gland extract, 91.6 per cent positive result was obtained with both male and female gland extracts. Common carp gland gave 66.6 per cent positive result with female gland extract and 55.5 per cent with male gland extract. Similarly, for silver carp, with both male and female glands 100 per cent positive results were obtained. Analysis of data indicated that differences recorded were not significant.

Introduction

The literature contains contradictory information concerning the gonadotrophic potency of pituitary glands from the male and female of the given species. Subsequent to the successful application of hypophysation in the thirties and the tremendous impact it had on the industry, several workers in Brazil, USA, USSR and in a few other countries hae attempted to find out whether any significant differences existed in the potencies of pituitary glands of male and female fish (Pickford and Atz., 1957); Clemens and Sneed, 1962). Fish culturists had come to use glands of either sex indiscriminately because they were not aware of any clear-cut differences among glands of male and female fish for successful hypophysation work (Fontenele, 1955). In the cat-fish *Clarias batrachus*, pituitaries of both male and female were equally potent and in *Heteropneustes fossilis*, the pituitary of sexually active males was less potent than that of females (Ramaswamy, 1962). Among amphibians, it has been reported that pituitaries from females are about twice as effective as those from males (Prosser, 1952).

After the first success in hypophysation of Indian carps in 1957, the methodology has been adopted throughout India for commercial production of carp spawn (Chaudhuri, 1967). Usually, while collecting glands from mature specimens, they are not segregated sex-wise.

In view of the differences expressed by several workers on this subject, experiments were carried out in 1966 and 1967 to find out if such differences existed in pituitaries of Indian carps. If any significant differences exist, the pituitaries must be separated by sex and used separately in order to achieve consistent results and to some degree establish and maintain uniformly in dosages in induced breeding work.

Material and Methods

Pituitary glands were collected from mature specimens of rohu (*Labeo rohita*), mrigal (*Cirrhinus mrigala*), common carp (*Cyprinus carpio*) and Chinese silver carp (*Hypophthalmichthys molitrix*) and preserved sex-wise in absolute alcohol. Several glands from different sexes were weighed in bulk independently and extracts prepared in distilled water and glycerine (Ibrahim and Chaudhuri, 1966), thus obtaining extract with uniform potency. The extracts were prepared and stored under identical conditions. Equal number of breeders were injected each time with equal dosages of either male or female gland extract. Rohu was selected as the recepient species. All breeders were treated similarly in handling and injections. As far as possible, fish of identical sizes and age were selected for each set of experiment. Females received a preliminary dose of pituitary extract of 2-3 mg and six hours later, an effective dose of 6-12 mg/kg body weight. Males were injected with one dose of 2-3 mg/kg body weight at the time females received the second dose. Usually the recipients spawned within 5-8 hours after the second injection. All the experiments were carried out in July-August in static pond water at Killa Fish Farm, Cuttack, with induced-bred and pond reared fishes.

Donor species	Gland extract – Female		Gland extract – Male	
	Number injected	Number positive & percent	Number injected	Number positive & percent
Rohu	37	28 (75.7)	37	28 (75.7)
Mrigal	12	11 (91.6)	12	11 (91.6)
Common Carp	9	6 (66.6)	9	6 (66.6)
Silver Carp	4	4 (100.0)	4	4 (100.0)
Total	62	49 (79.0)	62	49 (79.0)

A total of 124 sets of rohu were thus injected and the results obtained are presented below:

The results indicate that except in the case of rohu and common carp glands, differences in results could not be seen with mrigal and silver carp glands. Even the difference results with gland from males and females of rohu was not statiscally significant and the difference of 8.2 per cent positive result could be due to chance. Pooling of the positive results gives 79.0 per cent success with glands from females and 72.6 per cent with glands from males, a comparatively insignificant difference. The overall average was 75.8 per cent positive result. This is comparable to results (84.2 per cent) obtained by the authors in 1966 with non-sex segregated pituitary extract. In view of these results, it can be inferred that among these fishes, species as well as sex-specificity of pituitary, even if slight, do not prevent a high degree of positive results in induced breeding work.

Discussion

Obviously, the method of bioassay used in these experiments is relatively less elegant than the seminal thinning response technique of Clemens and Grant (1965), which might reveal greater differences. However, in regard to the sex-specificity of pituitary among Indian carps, the above experiments demonstrate that for successful hypophysation work sex and species segregation of glands is not essential. Pituitary from both male and female of these species as well as heteroplastic glands are equally effective. These findings are in agreement with the views of Clemens and Sneed (1962) who mentioned that although differences in gonadotrophic activity may exist according to sex of the donor or season in which collected, these difference appear to be insignificant in fish cultural practices when pituitaries are injected on a weight basis. In the same manner, even if differences in gonadotrophic activity exist among pituitary glands of male and female of Indian carps, these differences are not enough to seriously affect results.

[*Source: K. H. Ibrahim and Chaudhuri. H., I.J.F.S., December, 1970*]

Some Selected References

Chaudhuri, H., 1967. Breeding and selection of cultivated warm-water in Asia and Far East, *FAO Fish, Rep.*, (44) Vol. 4: 30-66.

Clemens, H.P. and Grant, F.B., 1965. The seminal thinning response of carp (*Cyprinus carpio*) and rainbow trout (*Salmo gairdneri*) after injections of pituitary extracts, *Copeia*, 1965 (2): 174-177.

Clemens, H.P. and Sneed, K.E., 1962. Bioassay and use of pituitary materials to spawn warm-water fishes. *Res. Rep., U.S. Fish Wildl. Serv.*, 61: 30.

Ibrahim, K. H. and Chaudhuri, H., 1966. Preservation of fish pituitary extract in glycerine for induced breeding of fish. *Indian J. Exp. Biol.*, 4(4): 249-250.

Pickford, G.E. and Atz, J.W., 1957. *The Physiology of the Pituitary Gland of Fishes.* New York Zoological Society, New York, 613 p.

Prosser, C.L. *et al.*, 1952. *Comparative Animal Physiology.* W. B. Saunders Company, Philadelphia and London.

Ramaswamy, L.S., 1962. Induction of spawning in catfish and frog hormones and an account of the cytochemistry of their pituitary glands. *Proc. Indian. Sci. Congr.*, 49(2): 133-151.

8. Carp Hatcheries Development

It precipates from the observations of Reddy that success of Induced Breeding initiated several fishery development processes which gains momentum over the next 50 years. Reddy observes, "In India, initially fish culture was started with the natural seed collected from the riverine system in the form of spawn and fry. In the course of time, induced breeding technique through hypophysation was developed and successful breeding of the Indian major carps and Chinese carps (Chaudhuri and Alikunhi, 1957; Chaudhuri, 1963; Alikunhi *et al.*, 1963, Chaudhuri *et al.*, 1966 Singh *et al.*, 1970) was achieved applying the technology. This method replaced the age-old practice of collection of carp seed from natural habitats (Chaudhuri and Singh, 1984).

After having achieved success in induced breeding of carps under captive conditions, a large number of hatcheries have been designed in the country for breeding and hatching carp eggs viz., Hatching pits, Chittagong type of hatching pits, breeding hapas, hatching hapas, floating hapas, earthern pot hatchery, transparent polyethylene sheet hatchery, galvanized iron jar hatchery, bin hatcheries, hanging dipnet hatchery, Chinese hatchery etc. While designing various hatcheries simultaneously, environmental conditions and other factors which are conductive for carp breeding have been investigated extensively in different parts of India (Hora, 1945, Khan, 1945, Chaudhuri, 1960, Dubey and Tuli, 1961, Alikunhi *et al.*, 1964). The studies indicated that dissolved oxygen, temperature, silt-free water and slow water current, and removal of metabolites are critical factors. In the above stated hatcheries, the factors which are conductive for breeding and hatching of carps have not been incorporated. These hatcheries mostly depended on the natural environment that varied from season to season, day to day and even within a day.

Carp Hatcheries Developed at CIFE, Mumbai

The Central Institute of Fisheries Education has been working on various aspects of carp hatcheries since its inception. Modification and modernisation of carp hatcheries have been initiated by S. N. Dwivedi who set up during 1976 a 6.35 litre capacity PVC vertical jar hatchery which was a replica of glass jar hatchery (Zoug jar Hatchery) (Bhowmick, 1974). Later, the 6.35 litre capacity PVC jar was modified into three versions of hatcheries made of galvanised iron C.G.I. i.e., 6.35 lit., 15 lit., and 20 lit. capacities during 1977 to 1979 by Dwivedi. These PVC and G.I. hatcheries were widely used in Andhra Pradesh, Maharashtra, Madhya Pradesh etc.

After having examined each of the essential factor responsible for breeding and hatching of carps through operation of above-mentioned hatcheries, the CIFE, Mumbai has taken the initiative of designing improved version of commercial hatcheries during 1980 to 1986. (Dwivedi and Ravindranathan, 1982; Dwivedi, *et al.*, 1983; Dwivedi and Reddy, 1986.

(Source: A. K. Reddy in Fishing Chimes: Jan–Feb, 2000)

Chapter 10

Composite Fish Culture, Pond Management and Record Fish Yield

A RESUME

This chapter presents two works of the Professor, one on "Pond Management in Composite Fish Culture" and another on "Composite Fish Culture and Record Fish Yield". The former one was published in 1971 and the latter in 1975 prior to the professor's posting in SEAFDEC. The field work in 1971 involved judicious section management, prestocking, fertilization, selection of species, density of stocking, harvesting and production in pond. The 1975 paper concerned research on Chinese Silver Carp and Grass Carp, Catla, Rohu, Mrigal and common carp which resulted in very high seed production in a small area. Use of well-balanced fish feed, judicious fertilization, elimination of the accumulated metabolites together with multiple cropping and periodic replenishment of the stock resulted in high production.

There are two very useful papers one publised in 'Aquaculture' in 'The Netherlands' and another unpublished. These concerned research on Chinese Silver Carp, Grass Carp, Catla, Rohu, Mrigal and common carp carried out over four years from 1971 to 1974 and on earlier works. The work by Chaudhuri, Chakraborti, Sen, Rao and Jena established that a very high fish production is possible with several species of fish in a small area. Use of well-balanced fish feed, judicious fertilization, elimination of the accumulated metabolites together with multiple cropping and periodic replenishment of the stock removed may make highly intensive fish culture in ponds with good yield of marketable fish possible in the future.

This chapter presents two works of the Professor, one on "Pond Management in Composite Fish Culture" and another on "Composite Fish Culture and Record Fish Yield". The former one was published in 1971 and the later in 1975 prior to the professor's posting in SEAFDEC. The field work in 1971 involved judicious section management, prestocking, fertilization, selection of species, density of stocking, harvesting and production in pond. The 1975 paper concerned research on Chinese Silver Carp and Grass Carp, Catla, Rohu, Mrigal and common carp which resulted in very high seed production in a small area. Use of well-balanced fish feed, judicious fertilization, elimination of the accumulated metabolites together with multiple cropping and periodic replenishment of the stock resulted in high production.

There are two very useful papers one publised in 'Aquaculture' in 'The Netherlands' and another unpublished. These concerned research on Chinese Silver Carp, Grass Carp, Catla, Rohu, Mrigal and common carp carried out over four years from 1971 to 1974 and on earlier works. The work by Chaudhuri, Chakraborti, Sen, Rao and Jena established that a very high fish production is possible with several species of fish in a small area. Use of well-balanced fish feed, judicious fertilization, elimination of the accumulated metabolites together with multiple cropping and periodic replenishment of the stock removed may make highly intensive fish culture in ponds with good yield of marketable fish possible in the future.

A. Stocking Pond Management in Composite Fish Culture (Paper Published in 1971)

Stocking pond management is the most important phase in composite fish culture which aims to obtaining maximum production of marketable fish per unit area of water. The main objective of the management technique is to take full advantage of the available pond space as also the utilization of all resources in a pond to produce fish. This method has come to be recognized as an effective way of intensive fish culture and is differently termed as mixed fish culture, multi-species culture, polyculture or association of fish species. This guiding principles in the rational management of pond in composite fish culture can be summed up as follows:

1. Culture of selected efficient pond fish.
2. Increasing the natural carrying capacity of the pond by proper fertilization and supplementary feeding.
3. Raising of compatible fish species with complimentary feeding habits.
4. Optimum utilization of the ecological niches of the pond by judicious stock manipulation.

Management of stocking ponds in composite fish culture involves a number of important steps which are enumerated below.

Selection of Pond

At a time when the pressure on agricultural land is steadily increasing it is desirable that such land is not used for fish culture. As a matter of fact, it is being emphasized all over the world that new land must be brought under intensive cultivation by utilizing soil not suitable for agriculture resulting in an increase in human food. The main criteria to be kept in mind while selecting pond site is that the soil is retentive and an adequate supply of water is assured.

A desirable pond is rectangular in shape with gradually sloping banks with an even bottom. It should have firm embankments and guarded inlets and outlets. Since bigger ponds are difficult to manage, area of ponds from quarter hectare to half a hectare can be considered to be within the ideal range. The depth of water in ponds would ordinarily to be 2 to 3 m and in no case be less than 1.5 m.

Pre-stocking Operations

Before a pond is stocked excess of aquatic vegetation has be removed. Some amount to be submerged rooted vegetation, however, may be left in the pond since these form food material for some carps on decay. These plants utilize nutrients bound up in pond soil and release them in the pond water thereby making the nutrients available for the growth fo algae. A small quantity of submerged rooted weeds are also required for providing 'cover' or "additional surface" for the attachment of fish food organisms. But the weeds should not be allowed to spread and hamper with fish movement or netting operations.

It is essential that predatory fishes such as the snake-headed murrels and the so called freshwater shark (*Wllago attu*) are removed.

Carp minnows and other weed fishes which compete for food with the cultivated species should also be eradicated. In practice, it has been seen that very often almost cent per cent of the fingerlings stocked survive till they are harvested, once the fish enemies are removed.

Of the various piscicides that are employed so far, mahua (*Madhuca latifolia*) oil cake has been found to be very useful since not only are the undesirable fishes eradicated but the oil cake also serves as a fertilizer resulting in production of zooplankters. A dose of 200 ppm is found to be effective. The distressed fish may be salvaged by repeated nettings. Poisoned waters get detoxified within two weeks. Alternatively any other fish poison of plant origin such as *Croton tiglium, Derris trifoliata, Milletia piscidia, M. pachycarpa* or *Barringtonia acutangula*, the bark or seed powder of which may be applied at 4-15 ppm depending on the plant used for clearing the pond of undesirable fishes. However, in absence of any fish poison of plant origin, chemical piscicide like 'Tafdrin–20' (20 per cent Endrin) at 0,01 ppm can be used. But since chemical fish posons are highly toxic and there is the danger of possible adverse effect of the persistant toxicity on the fish food organisms, these are not ordinarily recommended to fish farmers.

Pond Fertilization

It is well known that fertilization of fish ponds is an important means of intensifying fish culture. In India, in the traditional practice, fertilization of ponds is mainly confined to the nurseries. It is essential that pond fertilization should be extended also to the intensive fish culture programme. The fertilization schedule ought to be prepared after studying the quality of pond soil. For highly clayey soil very rich in organic matter inorganic fertilizer is better, whereas for highly sandy soil with low organic content organic fertilizer is recommended. A combination comprising both organic and inorganic fertilizers may be used for pond soil neither too clayey nor too sand with medium organic content. In ponds with acidic soil, pre-treatment with quick lime at 250-600 kg/ha may be recommended. In stocking ponds, fertilization programme should commence at least two to three weeks before stocking of fingerlings. This is followed by a number of instalments of fertilizers during the post-stocking period. In Cuttack ponds, a mixture of organic manure (Cowdung @20,000-25,000 kg/ha/yr in 7-11 instalments) and inorganic fertilizers (Am. sulphate/single superphosphate–cal.. amm. nitrate in the ratio 11:5:1, @1380-1725 kg/ha/yr in 4-10 instalments) have given encouraging results. Poultry droppings applied initially @5,000 kg/ha and then every quarter @1000 kg/ha and triple superphosphate @100 kg/ha every month is found to be very satisfactory. Organic manure when applied after introduction of fish should not be broadcast but put in baskets or dumped under water in the corners of the pond.

The manuring programme however is suitably modified depending on the growth rate of fish, available food-reserve in the pond and on its nutrient level, and also on the physico-chemical condition of pond water and climatic condition. An assessment of plankton both qualitatively and quantitatively gives a rough estimate of the available food reserve and analysis of soil and pond water may determine the nutrient level and physico-chemical condition of water.

Stocking

The stocking pond is generally stocked with fingerlings of sizes varying from 100-150 mm. In predator-free ponds the stocking size can be reduced to 75 mm to 100 mm. The fish stocked should be healthy and free from parasites and diseases.

Selection of Species and Species Combination

In intensive fish culture it is essential that important fish species selected should be fast growing, good converters of natural fish food into fish flesh and do not compete with each other for food. Keeping these points in view, the species found efficient for multi-species culture operations are the Indian major carps catla, rohu and mrigal, the Chinese grass and silver carp, and the Indonesian strain of common carp. These species are tolerant to each other and when put together for culture utilize the ecological niches in the pond efficiently to give higher yields. The functions of different species in the above combination are as follows:

(1) Catla and (2) silver carp are both surface feeders but while the former feeds mainly on zooplankters, the latter feeds predominantly on phytoplankton organisms. (3) Rohu is largely a column feeder feeding on a wide variety of algal forms, as well as decaying macrovegetation. (4) Grass carp is a voracious feeder on

certain types of macrovegetation; it occupies its own niche, does not compete with other species and contributes substantially towards the welfare of the surface and bottom feeders. The semi-digested excreta of grass carp are directly consumed by bottom feeders such as common carp and the decaying half-consumed vegetable matters also serves as a manure for plankton production. (5) Mrigal is a bottom dweller subsisting mainly on semi-rotten vegetable matters and detritus. (6) The common carp is an omnivorous scavanger, generally feeding at the bottom or at the pond margins in search of worms and other bottom organisms.

It may be observed from the foregoing account that such a combination would result in much better utilization of the pond space and the natural fish food it provides. It is, however, also observed that there is some amount of competition between the two surface feeders namely catla and silver carp and growth of catla is affected. But in spite of the competition, considering the high contribution of silver carp, the fastest growing fish we have, in mixed farming, it could be cultured together with the former by proper manipulation of the stocking ratio of these surface feeders. Similarly, although competition for food exists between mrigal and common carp, both the bottom feeders could be cultured together by putting them in proper proportion.

In addition to the above mentioned six important carps, some more suitable fairly quick growing species suitably fitting in the gap in the ecological niches may be selected. It is also essential to introduce a few predatory species whose size is such that the fingerlings of carps cannot be predated upon. These predatory fishes would keep the minnows and other weed fishes under control which are invariably present in a pond very often making their entrance through the inlets or outlets of even prepared stocking ponds. *Notopterus chitala* is a quick growing excellent table fish and grows to over 1.5 m in length. It is an ideal police fish which is worth trying in composite culture. *Mystus aor, Pangasius pangasius, Ompok* spp., *Channa punctatus* etc., can also be introduced in stocking ponds along with the cultivated carps.

In China, association of a large number of species is usually practiced. In Taiwan, in addition to the Chinese carps and common carp, mullet, tilapia, milk fish and also sea perch, a carnivore for controlling forage species are introduced. It is observed that tilapia alone has given as high as 2000-3000 kg/ha/yr production which is about half of the total production in mixed farming in Taiwan. Inclusion of tilapia in composite fish culture in India may be given a fair trial before rejecting the same totally.

The stocking combination as suggested above may, however, be sometimes modified according to the availability of the stocking materials of the various associated species, their adaptability to local climatic conditions and also according to the consumer preference in a particular region.

While from the general study of the food habits the ecologically different fish species are selected for composite culture, it is rather difficult to determine the optimum combination of the various cultivated species and there is very little experimental data available at present. For efficiently managing the pond resources it is not enough to select proper species for the fish association, but study should be made on a quantitative basis to find out the inter-relationship between and within the different fish groups in ponds and the optimum number to be stocked under different environmental conditions should be worked out (Yashouv, 1968). Experiments on composite fish culture conducted at the Cuttack Substation have given satisfactory results with various species combination of catla 10-12.5, silver carp 20-25, rohu 25-30, grass carp 10, mrigal 12.5, common carp12.5 and calbasu or gourami 2.5 in terms of percentage. However, further experiments have to be carried out to arrive at the optimum ratio of various species to obtain the maximum possible yields.

Stocking Density

There is no information based on experimental data as to the optimum density of stocking of fishes in intensive fish culture in India. The stocking density is primarily dependent on the natural fertility of a pond but the same can be considerably increased by stocking combination of several compatible species and increasing the carrying capacity of the pond by proper fertilization and supplementary feeding. At Cuttack, a stocking rate of 3750/ha of Indian major carps could be increased to 5000/ha by association with the exotic Chinese carps and common carp. In Taiwan, the stocking density in multi-culture is as high as 9730 to 16150/ha. Ordinarily stocking rate is calculated and expressed according to the surface area of a pond. But as the volume of water is of great importance in intensive fish culture for the diluting effect of the growth inhibiting factors like body excretions of the fish it is advisable that stocking density is determined in terms of hectare meter.

Post-stocking Operations

Among the various post-stocking operations, supplementary feeding is the most important. By adoption of supplementary feeding a larger stock can be maintained in a pond with more efficient utilization of the natural fish food organisms. Artificial feeding with carbohydrate-rich cereal wastes in association with oil cakes fairly rich in protein contents is resorted to after the fingerlings are stocked. Commonly used supplementary feeds are 1:1 ratio of mustard/ ground nut oil cake and rice bran/wheat short. The quantity broadcast per day, generally in the early hours of the morning, is 5-10 per cent of the body weight of the fish stocked. Mixing of chopped pieces of green vegetations along with the feeds is also done. It is essential that a cheap and balanced feed with high conversion value be evolved and made available to the fish culturist for enhancing the production of fish in his pond. Although grass carp requires additional feeds like submerged aquatic weeds, duckweeds or land grasses, it can also be fed with rice bran and oil cakes mixed with chopped green land vegetation. The quantity of the supplementary feed to be provided is sometimes very difficult to determine. It is advisable that a fish farmer should himself determine the quantity of feed by actual observation in the field. When the fish are able to consume all the feed supplied, the feeding schedule should be increased keeping in view that there is no waste. Unconsumed feed may develop a thick bloom which is undesirable and may lead to mortality of fish especially on hot summer days.

To check the growth of the fish it is desirable that these are sampled once a month or at least once in two months. Fertilization and feeding schedules may be altered based on the performance of the fish as evidenced by such periodical samplings.

Harvesting

Though there is hardly any experimental data as to the comparative merits of rearing fish for a period of 1 yr or longer, it is advisable to thin the population as soon as the fishes reach marketable size. Marketable or table size also varies from species to species. For Indian and Chinese carps it is rational to take one kg fish as the marketable size. Fishes attaining 1 kg size may be removed and marketed and fingerlings of the same species may be replenished. Harvesting can be profitably done in summer when the water level is very low or in monsoon when the selling price of fish is the highest. A fish farmer has to use his own judgement from his experience as to the proper time he should completely fish out his ponds.

Production

There is no doubt that composite culture of a number of compatible species of fish increase production considerably under proper management technique. In the experiments carried out at the Cuttack Substation with combination of three Indian major carps catla, rohu and mrigal, an average yield of about 2000 kg/ha/yr has been achieved, when the known average production in the country is only 600 kg/ha/yr. Similar experiments when conducted with exotic carps (grass carp, silver carp and common carp) alone the average production was raised to about 2900 kg/ha/yr. But when both the Indian and exotic carps were included in the composite/ culture, the highest production of 4210 kg/ha/yr was achieved. In Taiwan, the highest yield recorded so far has been 6960 kg/ha/yr. In Burma, the maximum production obtained from a pond exceeded 10,000 kg/ha/yr. It is expected that with judicious selection of ecologically different species of efficient pond fish and determination of proper combination and stocking density, with more efficient supplementary feeding and proper fertilization programme, introduction of small predators and eradication of weed fishes, continuous harvesting of marketable fishes and stock manipulation, the overall production of fish by composite culture can be increased manifold. The economics of production, however, should receive due consideration in the management of composite fish culture.

B. Paper Published in 1975

The revolutionary success in record yields of fish production by composite fish culture in freshwater ponds was published in Aquaculture (1975) 343-356 from 'The Netherlands'. It was a collective work by H. Chaudhuri, R. D. Chakraborty, P. R. Sen, N. G. S. Rao and S. Jena of Central Inland Fisheries Research Sub-station, Cuttack, Orissa.

Prof. Chaudhuri and his co-workers utilized this known how in their work while on deputation to the countries in South, Southern Asia and else where to augment fish yield in those countries.

An abstract of the publication goes as follows:

Composite fish culture of the Indian major carp, *Catla catla, Labeo rohita* and *Cirrhinus mrigala* together with the Chinese silver carp, *Hypophthalmichthys molitrix* and grass carp *Ctenopharyngodon idella* and the Indonesian strain of common carp (*Cyprinus carpio communis*) was carried out in experimental ponds in 1971–1972 and 1973-1974. The management techniques adopted aimed at high yields of marketable fish (around 1 kg) in 1 year.

In the first year, production varied from 2889 to 5600 kg/ha/year and this rose to an average of 8200 kg/ha/year with a maximum of 9389 kg/ha/year in the second year. The important changes made to the management schedule in 1973 were increased stocking density, greater use of feed of fertilizers, and provision of more weeds for grass carp.

The experimental studies demonstrate the practicability of raising large crops of healthy marketable fish in India.

Introduction

Composite fish culture or polyculture of indigenous carp together with the Chinese carp and the common carp has been found to result in high fish production in freshwater ponds in India. Lakshmanan *et al.* (1973) and Chaudhuri *et al.* (1974) have reported experimental studies on the different forms of composite fish culture tried in India, the highest production recorded in these studies being 7500 kg/ha/year (Chaudhuri *et al.*, 1974). The present state of development of the technique has been reached after a decade of applied research (Chaudhuri and Chakraborty, 1974) and marks the beginning of improved technology in India which results in fish yields of over 5000 kg/ha/year approaching those reported from some countries in South-East Asia, China and Israel (Rabanal, 1968; Bardach *et al.*, 1972; Yashouv and Halevy, 1972).

The experiments were conducted to increase fish production per ha adopting improved management techniques aimed to producing.

Experiments Undertaken in 1971–1972

The weed-free, experimental ponds were first poisoned with mahua (*Madhuka latifolia*) oilcake for complete removal of the previous stock of fishes and then stocked at 5 000 fingerlings/ha in December 1971. The stocking material comprised the three Indian major carp: catla, *Catla catla* (C), rohu, *Labeo rohita* (R) and mrigal, *Cirrhinus mrigala* (M), together with the Chinese silver carp, *Hypophthalmichthys molitrix* (SC) and grass carp, *Ctenopharyngodon idella* (GC), and the Indonesian strain of common carp, *Cyprinus carpio corn-munis* (CC) in the ratio 1:3:2:2:1:1. In F.P.2 the ratio of mrigal and common carp was different – 1.6:1.5 instead of 2:1. A certain number of carp hybrids (*Labeo calbasu* male x *Catla catla* female) and grey mullet (*Mugil cephalus*) were also introduced to each pond raising the stocking density from 5000 to 5175–5334/ha (Table 1). The ponds were harvested in December 1972 after a complete year.

Organic (cowdung) and chemical fertilizers (urea and triple superphosphate) were used alternately to fertilise the ponds. The quantity of cowdung used was a little over 4.6 tonnes/ha/year in R.P.12, 13, and 14 while in F.P.2 it was a little over 3.7 tonnes/ha/year. Urea was used at a rate of 72 kg/ha/year and triple superphosphate at 55 kg/ha/year.

The fish were fed each day with a mixture (1:1 by weight) of ground-nut oilcake and rice bran. The feed was broadcast over the water surface, near the pond margin at particular spots. The quantity of feed used was 5155 kg/ha/year in each of ponds R.P.12,13 and 14, and 3632 kg/ha/year in F.P.2. Application of fertilizers and feed was reduced and, at times, suspended during the summer months when the water level was low and algal blooms developed.

Aquatic weeds, mostly *Spirodela* and *Hydrilla* and occasionally *Ceratophyllum* and *Najas*, were provided as feed for the grass carp. Approximately 7.5 tonnes of weeds/ha/year were fed to the grass carp in each of the ponds.

In R.P.12,13 and 14, intermittent harvesting was carried out and an equal number of catla fingerlings were replenished in R.P.13 and 14. F.P.2 was harvested only once, at the end of 1 year.

Data on the plankton content and nutrient level of ponds were collected once a month. Length and weight measurements were also recorded once a month, except during the peak summer period to avoid mortality due to handling. About 30–50 fish of each species were measured and weighed. At the end of the experiment, the fish were harvested by netting and, those that remained, by application of mahua oilcake at 200 ppm. The entire catch was weighed one species at a time. The average weight attained by each species and its percentage contribution to total production as well as the production /ha were computed.

Results

Physico-chemical Conditions of Water

The physico-chemical conditions of the water of the different ponds are given in Table 1.

The increase in the average nitrate level in all the ponds during 1973–1974, despite the greater use of nitrates for growth is to be associated with the application of larger quantities of inorganic fertilizers and feed, both supplying nitrogen. The ready utilization of phosphate by plankton would perhaps account for its low average values in 1973–1974.

Table 1
Physico-chemical Conditions of Water in the Experimental Ponds

	R.P.12	R.P.13	R.P.14	F.P.2
1971-1972				
pH	8.2	8.2	8.2	8.2
Dissolved oxygen (ppm)	5.4	6.4	6.5	6.1
Total alkalinity (ppm)	121.5	119.1	118.3	128.7
Nitrate (ppm)	0.241	0.260	0.239	0.429
Phosphate (P_2O_5) (ppm)	2.426	2.10	1.95	1.77
1973-1974				
pH	7.8	7.8	7.86	7.81
Dissolved oxygen (ppm)	6.99	7.1	6.71	6.86
Total alkalinity (ppm)	158.2	155.6	131.6	130.1
Nitrate (ppm)	0.754	0.862	0.941	0.547
Phosphate (P_2O_5) (ppm)	0.51	1.103	1.36	1.14

The water temperature varied from 22.0 to 35.0°C in 1971–1972 and 22.2°C to 39.0°C in 1973-1974.

Plankton

The average plankton density in R.P.12,13 and 14 and F.P.2 was low, being 0.15, 0.17, 0.22 and 0.36 cc/45 1 respectively in 1971–1972 and 0.31, 0.21, 0.35 and 0.46 cc/45 1 respectively in 1973–1974. Among phytoplankton, *Scenedesmus, Pediastrum, Merismopedia, Navicula, Closterium, Oscillatoria* and *Tetraspora* were more prominent, whereas dominant zooplankton were rotifers (mostly represented by *Brachionus, Keratella, Asplanchna,* and *Filinia)*, nauplii of copepods, *Cyclops* and *Moina*.

Growth and Survival

Catla and silver carp (C 1: SC 2), which together accounted for 30 per cent of the numbers stocked in 1971–1972 contributed between 37 and 40 per cent of the production. In 1973–1974, these two species (C 1: SC 2.5) formed 35 per cent of the numbers stocked but produced between 37 and 41 per cent of the total crop, which is about the same as in the previous year. The bottom feeders mrigal and common carp (M 1.5: CC 1.6) together formed 30 per cent of the numbers introduced, but in terms of production accounted for 25 to 28 per cent of the harvest in 1971–1972, but did somewhat better in 1973–1974 when their contribution to the harvest ranged between 26 and 31 per cent (M 1: CC 2). Grass carp, which formed 10 per cent of the stocked fish in both years, formed 9 to 14 per cent of the yield in 1971–1972, but its contribution increased to 10 to 19 per cent in 1973–1974. The percentage contribution of rohu to total production was invariably less than the percentage introduced. The species attained an average weight of over 1 kg, while its survival was comparatively low.

The average monthly increment of the different species is given in Table 2.

Yashouv and Halevy (1972) recorded daily weight gains of 6.6–6.9 g in silver carp in ponds fertilized and provided with supplementary feed of protein-containing pellets when the density was 1 500 fish/ha. At the higher density of 1 875/ha, silver carp recorded a daily average weight gain of over 4 g in our ' 1973–1974 experiments.

Catla male x rohu female hybrids had a better growth rate than the hybrids of calbasu male x catla female. The net increment in weight of over 1 kg in a year by chital *(N. chitala)* in R.P.12 and 13 shows the value of the species to fish culture.

Table 2
Average Monthly Weight Increment of the Indigenous and Exotic Carps

Species	1971–1972* (g)	1973–1974* (g)
Silver carp	124.70	124.61
Catla	86.42	81.02
Rohu	76.62	70.73
Grass carp	140.51	174.77
Mrigal	62.82	95.46
Common carp	148.01	98.79

*R.P. 14 not included.

The survival percentages were satisfactory, except in the case of R.P.14 in 1971–1972. The poorer survival rate of fish in R.P.14 was on account of loss due to poaching. The pond F.P.2, away from the periphery of the farm, was not easily accessible to poachers, hence the high survival of fish in this pond in both years.

Fish Production

The gross and net production recorded for each pond in 1971–1972 and 1973–1974 are given in two tables. The production during 1973–1974 may be considered a record yield, as the average net production for the four ponds was over 8 200 kg/ha/year which is a new Indian record in fish production from ponds and comparable to the best recorded elsewhere.

Discussion

The most outstanding feature of these experiments was the attainment of marketable size (around 1 kg) by the majority of the species. Attainment of an average weight of over 1.5 kg by silver carp (over 95 per cent survival), between 1 and 2 kg by common carp (over 95 per cent survival) and over 1 kg by mrigal (100 per cent survival) (Tables 1 and 2) are notable achievements and better than those obtained so far for these species in composite culture experiments in India.

Though the production of 9 389 kg/ha/year obtained in F.P.2 in 1973–1974 can be considered the highest fish production achieved in freshwater ponds in India, the production of 8 837 kg/ha in 286 days of culture in R.P.13 can perhaps be taken as an even greater achievement. A lesser quantity of feed was, however, used to raise the fish in F.P.2 which is consequently more economical.

It may be remembered that more than three times the quantity of weed used in R.P.13 was used in F.P.2 in 1973–1974. The beneficial effect derived by other fish species as a result of feeding weeds to grass carp has been noted in earlier work on composite culture. The particular beneficiaries seem to be the bottom and marginal feeding common carp and the iliophage mrigal whose performance was much better in F.P.2 in 1973–1974, where grass carp also had a much better growth rate. This can also be seen from the performance of these three species in 1971–1972.

Of the two sizes of silver carp released in R.P.12, 13 and 14 in 1973–1974 (Table), the larger ones attained marketable size in 3 months. These fishes were removed from R.P.12 and 14, even so, all the fish in the replenished stock attained marketable sizes within the remaining period of culture. Similarly, additional numbers of catla fingerlings, introduced 3 months after the start of the experiment to facilitate phased harvesting in R.P.12 and 14, also attained marketable sizes.

In 1971–1972 the highest production was achieved in R.P.13 which was periodically harvested, but in 1973–1974 both F.P.2 and R.P.13, harvested at the end of the culture period, produced more than the other ponds which were periodically harvested. It can be seen from Tables 1 and 2 that in multi-species culture rohu does not normally attain a weight of 1 kg or more when it comprises 25–30 per cent by number of the stocked fish and a lower rate of stocking (15–20 per cent) may be more useful. However, in ponds from which marketable fish of the other species were removed in large numbers early in the rearing period, much better growth of rohu was obtained, even when stocked at a higher rate (Chaudhuri *et al.*, 1974).

Despite an increase in the stocking density from 5 000/ha in 1971–1972 to 7 500/ha in 1973–1974, common carp, rohu and silver carp performed as well, due to improved management. Mrigal did even better and catla and grass carp performed only slightly worse. This shows the vital role played by the increased quantity of feed provided during 1973–1974. The increase in the species ratio of silver carp and common carp did not deter the successful performance of these species but the larger number of silver carp would appear to have somewhat adversely affected the growth of catla in 1973– 1974 (SC:C = 2.5:1).

Another interesting feature was that the average weight of the females of mrigal was greater than that of the males of the species. The females were in the third stage of maturity but were more than 1.5 times as heavy as the males. This striking difference suggests a faster rate of growth of the females. A similar tendency was also noticeable in silver carp. This could not be ascertained in case of the other species, because of lack of external evidence of sexual dimorphism at the time of harvesting.

N. chitala may be considered as a desirable component of the species in composite fish culture since it takes care of the young of common carp which are produced as a result of "wild" spawning and also small prawns, insects and extraneous minnows that may enter the pond. This appears to be the first record of the performance of the species under systematic fish culture. The performance of the grey mullet *M. cephalus* can be said to indicate the possibility of its use in polyculture in freshwater ponds in India. Yashouv (1972) mentions the excellent performance of the species in polyculture experiments in carp ponds in Israel.

Conclusions

During the course of these experiments, both feeding and fertilization had to be curtailed or stopped altogether on a number of occasions because of development of algal blooms, mostly *Microcystis*, which cause depletion of oxygen and fish kills in the early hours of the morning. During such periods, it was also observed that the fish were being increasingly infected by the fish louse *Argulus*. Probably, replenishment of water loss during culture or its reuse after purification would further increase fish production from these ponds by controlling the algal blooms, reducing the extent of infection from the fish and increasing the oxygen content. The use of compressed air from perforated tubes on the bottom of the pond in fish culture under super-intensive methods has been reported by Marek and Sarig (1971).

Non-availability of desired aquatic weeds for grass carp can greatly hamper the performance of the species, whereas phenomenal growth can be obtained by providing it with large quantities of such weeds. However, the introduction of weeds from outside must be done with caution lest it serve as a source of infection.

The new high in fish production in India achieved by composite culture of indigenous and exotic species in experimental studies has amply demonstrated the practicability of raising large crops of healthy marketable fish there. Use of well balanced fish feed, judicious fertilization, elimination of the accumulated metabolites together with multiple cropping and periodic replenishment of the stock removed, may make highly intensive fish culture in ponds with good yields of marketable fish possible in the future.

Some Selected References

Chakrabarty, R.D., Murty, D.S., Sen, P.R., Nandy, A.C. and Chakrabarty, D.P., 1972 a. Periodic harvesting to enhance production in composite culture of Indian and exotic carps. In: *Silver Jubilee Symp., Aquaculture as an Industry,* Centr. Inland Fish. Inst. Barrackpore, p.27 (Abstract).

Chakrabarty, R.D., Murty, D.S., Sen, P.R., Nandy, A.C. and Chakrabarty, D.P., 1972 b. Short term rearing of Indian and exotic carps. In: *Silver Jubilee Symp., Aquaculture as an Industry,* Centr. Inland Fish. Res. Inst., Barrackpore, p.25 (Abstract).

Chakrabarty, R.D., Sen, P.R. and Rao, N.G.S. (In Press). Harvesting of ponds under composite culture by netting–species representation in the hauls.

Chaudhuri, H. and Chakrabarty, R.D., 1974. Intensive Fish Farming and Development of Technique of Composite Culture of Indian and Exotic Fishes. *Lecture, Summer Inst. Inland Aquaculture,* Cuttack, organised by Centr. Inland Fish. Res. Inst., Barrackpore (ICAR), 10 pp., mimeographed.

Chaudhuri, H., Chakrabarty, R.D., Rao, N.G.S., Janakiram, K., Chatterjee, D.K. and Jena, B., 1974. Record fish production with intensive culture of Indian and exotic carps. *Curr. Sci.,* 43(10): 303–304.

Singh, S.B., Sukumaran, K.K., Chakrabarti, P.C. and Bagchi, M.M., 1972. Observations on composite culture of exotic carps. *J. Inland Fish. Soc. India,* 4: 38–50.

Chapter 11

Some Selected Works of Professor Chaudhuri

A RESUME

This chapter records the contents of the stray papers recovered from the godown of Professor Chaudhuri. These are on a few subjects the contents which are presented in brief. The subjects are as follows:

1. Culture of air breathing fish.

2. Rice and Fish culture.

3. Role of grass carp in control of weeds.

This chapter records the contents of the stray papers recovered from the almirah's of Professor Chaudhuri. The subjects are as follows:

1. Culture of air breathing fish.

2. Rice and Fish culture.

3. Role of grass carp in control of weeds.

1. Culture of Air Breathing Fish

Fish culture aimed at supplying valuable protein to human diet has been receiving due attention in India in recent years. For increasing fish production all available culturable waters are required to be brought under scientific fish culture. The readily available cultivable freshwater areas could be suitably utilised for the culture of quick growing cultivated species such as Indian and Chinese carps. But there are other water areas still to be reclaimed which on a moderate estimate constitute to about 2 lakh ha of swampy and marshy areas. The reclamation cost for making these areas suitable for carp culture may be prohibitive. In view of this it is desirable to utilise these swampy and derelict waters for culturing the economically important air-breathing fishes such as, the so-called climbing perch, koi (*Anabas testudineus*), the catfishes, magur (*Clarias batrachus*) and singhi (*Heteropneustes fossilis*) and the snake-headed murrels (*Channa* spp.) which can thrive fairly well in swampy and in oxygen depleted foul waters.

These air-breathing fishes are also termed "live fishes" because they are marketed alive. They possess accessory respiratory organs by means of which they are capable of utilizing oxygen directly from the atmosphere for respiration. The air-breathing fishes, especially koi, magur and singhi have high demand in some parts of India, fetching a relatively high price because they are believed to have highly nutritive recuperative and medicinal values and considered as ideal food for sickly and convalescing people. The same is also true in a few South-East Asian countries where these live fishes are indigenous. Murrels are very popular and are considered as excellent table fish in peninsular India and also in Punjab, Madhya Pradesh and a few other States. By culturing air-breathing fishes the unhealthy surroundings of marshy areas can be converted to productive fish farms. Since these fishes are insectivorous in habit either during some stages of the life-cycle or throughout life, they may also curb breeding of mosquitoes in such waters. The live fishes account for 9.8 per cent of the marketable surplus of fish in India.

Live fish culture is still in its infancy. Although studies on the life-history, bionomics, feeding and breeding habits of majority of the common air-breathing fishes have been made, very little information is however, available on the techniques of culture of these fishes. In India, no well organized scientific method of cultivation of these fishes is practised at present. In the absence of any well established traditional methods, or culture methods evolved from scientific experimentation, some plausible suggestions could only be made with the help of the present knowledge of their bionomics, feeding and breeding habits. Information on their natural food at various stages, growth, size at maturity and breeding habit are of great importance in the successful cultivation of these air-breathing fishes.

Koi, Singhi and Magur

The food and feeding habits of koi, singhi and magur are more or less similar. The larvae and young fry feed mainly on phyto- and zooplankters including micro-crustaceans and also insects. The juveniles and adult feed on insects and insect larvae, crustaceans, shrimps, ostracods, worms and also on algae, higher plants and organic debris. They are predominantly insectivorous and though predatory in habit are not markedly piscivorous.

Koi grows to about 12 cm in one year and attains maturity at 8 cm size. The breeding season is during monsoon. Eggs are very small, almost transparent, pelagic and hatch out in 24 hours at 27°-30° C. The species breeds in rivers, in paddy fields and also in seasonal ponds. They do not usually breed in perennial stagnant water but migrate from the pond to the nearby fields during rains for spawning. Sometimes they are observed to migrate to prepared carp nursery ponds and breed there. The fish can be bred even in aquarium in collected rain water and in the laboratory by hormone injection.

The air-breathing catfishes magur and singhi breed in ponds almost throughout the year with a peak spawning period during monsoon months. They attain maturity in the first year and attain a size of 20 cm in 1 year. Magur and singhi breed in ponds and very often migrate during rains to adjacent fields for spawning. Eggs are attached to grass or any marginal or aquatic weeds. Singhi eggs are greenish in colour whereas the colour of magur eggs is yellowish-brown. The incubation period is 20-24 hours at a temperature of 27°-30° C. In Thailand magur has been observed to spawn inside a hole prepared by the female. The eggs are deposited in the hole adhering to the grass or sometimes to the soil surface. The fish is usually bred in specially prepared breeding pond provided with horizontal holes (20cm x 35 cm) dug on the pond bank at about 20-25 cm below the water surface. The eggs are collected from the holes and put in troughs for hatching. Fry are reared in wooden tanks. About 2200-5400 fry are obtained from one couple. Singhi and magur can be induced to breed very easily by hormone injections also.

Since koi, singhi and magur mainly feed on insects, worms and bottom biota, these can be cultured with advantage in ponds with some amount of submerged vegetation as the weeds would harbour insects. The stocking materials may be obtained by induced breeding, or by providing proper breeding facilities in ponds or in the laboratory aquaria or collected from the natural waters. All the three species can be cultivated together in suitable proportions. Heavy manuring with cheap, readily available organic manure such as cowdung, pigdung or compost to encourage production of worms, insects and bottom organisms may be done. Introduction of the small prawn species *Macrobrachium lamarrei* in the ponds would provide food for these air-breathing fishes. Artificial feed such as, rice, rice bran, fish meal, chopped fish, meat and vegetables may be provided, for these fishes. The conversion ratio of the feeds may also be worked out for the assessment of their economics. Sidthimunka *et al.* (1968) have given an account of the culture of magur in Thailand giving very high yields. According to them most successful farm yielded 107,500 kg/ha fish from 656,250 kg of feed (conversion ratio 6:1). The optimum stocking rate was 50 fingerlings/m² pond surface. The fingerlings were fed with chopped meat or fish offal mixed with rice, vegetable and peanut cake. The average growth is 40 g/month and the marketable size is reached in about 4 months indicating that 2-3 crops can be raised in a year. Very high population of singhi and magur have been observed in pagoda ponds in Burma where these fishes are fed heavily by the visitors with bloated rice and bundles of Ipomea aquatic a leaves and stem.

Murrels

The murrels, especially the larger species, the striped murrel (*Channa striatus*) and the giant murrel (*C. marulius*) are highly predatory and piscivorous. They feed on fish, frogs, insects and other live food. The smaller species, the spotted murrel (*C. punctatus*) feeds mainly on minnows and other small fishes, shrimps, insects and occasionally on molluscs. The early fry of murrel feed on zooplankters and the fingerlings feed largely on insect, insect-larvae, shrimps and on fish fry.

The murrels breed in ponds almost throughout the year the peak period being immediately preceding and during the rainy season. Sexual maturity is attained in the striped murrel (maximum size–90 cm) in 1 year when the fish reaches about 25 to 30 cm size. It may even grow up to 48 cm in the first year under favourable conditions. The giant murrel (maximum size–1.2 m) is a quicker growing one and is reported to attain a maximum size of about 75 cm at the end of the first year. The spotted murrel (maximum size–30 cm) becomes sexually mature in the first year. The murrels make a sort of nest by clearing a small area among shallow marginal weeds. But they are observed to breed even in pond devoid of any vegetation. The eggs are pelagic, pale yellow or amber coloured and each has a single large oil glouble. The incubation period ranges from 24-40 hours. Both the parents guard the nest and also the brood of fry when they move about in shoals. The young ones of different murrel species can be recognised from their conspicuous colour. The fry and fingerlings of the giant murrel have an orange yellow longitudinal band on each side of the body and an ocellus at the upper half of the base of the tail whereas those of striped murrel can be distinguished from their vermillion red body colour. Young ones of the spotted murrel have lateral bright golden yellow band on a blackish ground body colour.

Although there is no well organized culture of murrels, some experimental culture was carried out in Bombay, Punjab and the erstwhile province of Madras. These, however, did not give any encouraging results. The survival of fingerlings was very low. Though murrels breed readily in ponds, hatching of eggs in the laboratory and rearing of larvae even with aeration were found to be difficult (Alikunhi, 1957). So collection of fry from nature for stocking purposes is better to begin with. The conspicuously coloured murrel fry which move in shoals on the surface of pond water could easily be located and collected by a piece of net. Well prepared nurseries with abundance of zooplankton are suitable for rearing larvae as well as early fry to fingerling stage. When the fingerlings are stocked in larger stocking ponds, a steady supply of forage fish is to be ensured. In the preparation of stocking ponds, some forage fish such as, minor barbells (*Puntius spp.*), *Amblypharyngodon*, *Esomus*, *Oxygaster*, *Ambassis*, etc., should be stocked sufficiently early so that the forage species could breed and sufficient number of their young ones of 12-25 mm. size are available to be fed upon by murrel fingerlings. A steady supply of forage fish of suitable sizes for predation by the piscivorous species is to be ensured for successful culture of the latter species. The forage species should be so selected that it breeds freely and prolifically in ponds, grows quicker, attains early maturity and that the maximum size is such that it cannot be predated upon by the piscivorous murrels. By selecting an ideal forage fish determining the stocking rate, and by manipulation of forage-predator ratio, a balanced population can be maintained which would provide adequate number of forage fish and at the same time prevent their over population. At present no definite information on the forage fish-carnivorous fish (F/C) ratio is available in respect to the predatory fishes in India.

Experiments carried out at Cuttack on the culture of forage fishes like, *Puntius sophore* and *Oxygaster bacaila* have shown that they breed prolifically in ponds and just 20 pairs could give a production of 472 kg/ha in five months (Alikunhi, 1957). *Tilapia mossambica* seems to be an ideal forage fish to be cultured along with murrels because of its prolific breeding, better parental care, hardiness, omnivorous feeding habit, attaining maturity early and also growing to a fairly big size. Its culture with a predatory species may lead to proper control of its population by predation so that the fishes which escape predation grow to marketable size and give high production.

In an experiment in Thailand (Tongsanga, 1962) in which *C. striatus* and tilapia were cultured together, the former gave a recovery percentage of 53.7 while the latter proliferated to more than 22 times the original number stocked within a rearing period of 290 days. The total production of harvestable fishes was much better than culture of tilapia alone. Preliminary experiments conducted at Cuttack with tilapia and the striped murrel, however, did not give very encouraging results. Tilapia was also tried as forage fish in Madras (Arumugam, 1966). But this did not prove successful. It was suggested that tilapia being a fish with compressed body and spiny fins is not easily consumable by murrels.

From the above account it appears that some of the soft bodied forage fish such as, minnows, minor barbels, etc., would provide better food for the murrel species.

It may, however, be suggested that for murrel culture, the stocking ponds should be manured with cowdung or poultry droppings or compost for production of food for the forage fish. The pond may have marginal weeds far harbouring aquatic insects which are important food items of fingerlings of murrels. Breeders of selected forage fish (500-600 pairs/ha) should be stocked one to two months before the monsoon–the peak breeding season of the species. Murrel fingerlings 8-12cm long either collected from natural waters or reared in nursery ponds should be released 3-4 months later at 3,000–4,500/

ha. Artificial feed in the form of rice, rice bran, oil cake and also slaughter house refuse or chopped cheap trash marine fishes may be provided.

Harvesting of the air-breathing fishes may pose some problems. Since they are bottom dwellers and adept in avoiding net, it will be very difficult to catch all the fish by netting. Setting of nets at the bottom and various types of traps can be improvised for removing the stock. In Thailand, harvesting of magur is done by partial drawing down of pond water and fishes captured by seines and scoop nets. Investigations, however, should be carried out to evolve suitable gears to catch these air-breathing fishes easily and effectively. Although poisoning may be resorted to in experimental culture for obtaining accurate data, it cannot be recommended as a regular practice since the entire population of forage fishes will be wiped out and it will take a long time for them to reestablish in the pond to provide food for the piscivorous species. This would work against the economics.

The failure in the culture of live fishes very often may occur due to the migratory habit of these fishes. They can easily migrate over moist land during or after a rain fall. To prevent migration some low fencing a foot or so in height should be provided around the stocking pond. The inner side of the bundh in magur ponds in Thailand are made very firm by pressing with heavy log so that the fishes cannot climb over or burrow through the bundh and escape. Fencing with bamboo or wire net to a height of 50 cm is provided around the pond; in some cases wooden planks are used.

From the foregoing account it is observed that live fish culture is till in the experimental stage in India. But considering its importance and potentialities in the development of fisheries in India the culture of live fish should receive adequate attention. It is expected that with the evolution of a proper culture technique, an increased production of fish could be obtained also from swamps and other derelict waters which are unsuitable for carp culture.

2. Rice and Fish Culture

Introduction

Asia, where over 90 per cent of all rice produced in the world is grown and consumed has about 85 million hectares of irrigated rice fields that are suitable for rice-fish culture. Yet most of these fields remain untapped for aquaculture.

Rice and fish form an ideal staple food for most Asians. Rice is the main dietary source of carbohydrates while fish, being a source of cheap and easily digestible protein, fills a significant part of, or supplements, the protein requirement. Rice-fish culture is a centuries-old practice. Although quite low, the fish yield obtained from traditional rice-fish culture provides additional income in rural areas. Until the late fifties, several countries raised fish in rice fields by modifying the fields to provide more water depth for the fish. With the introduction of high yielding rice varieties, however, and increased use of pesticides which are mostly highly toxic to fish, the fish production in rice fields declined. Also, adoption of multiple cropping system for rice and shortening of cropping duration left, limited scope for fish culture in rice fields because the fish rearing period was cut drastically.

Integration of fishfarming with crop and livestock production systems has received worldwide attention only in recent years. More recently, renewed interest arose with the success in developing pest-resistant rice strains and the discovery of insecticides which are better adapted for use in rice-fish culture. Also, many farmers have adopted rotation of rice and fish crops, which largely eliminate the hazards of the harmful effects of pesticides on fish.

In this article an attempt has been made to collate briefly available information on the past and present status of aquaculture in rice fields. Till the mid-fifties very little, reliable information and literature was available on rice-fish culture. It was through the effort of FAO in Rome that interest was generated, and quite a number of useful publications came out and several reviews made on rice-fish culture. This article also briefly mentions the scope of future development of this integrated system with advance in research of rice and fish culture.

Integrated Rice-Fish Production System

Integrated rich-fish production system may be classified into two categories:

A. Captural System

In this system there is no selective stocking of fish. The fish seed from the wild enter the rice-fields, grow there and are harvested along with the rice.

B.   Cultural System

In this system the rice field is deliberately stocked either by making devices allowing the entry of the fish from the wild or by stocking definite number of seed of selected species. Inputs are provided for proper water management: constructing trenches, sumps and sluice gate and also in some cases providing fertilizers and supplementary feed. This system again is of two types:

Simultaneous Culture of Rice and Fish

Rice and fish are cultured in the same field, rice being the main crop and fish as the secondary one. It has the following advantages:

☆ Culture of fish is usually beneficial to rice. The grain output increases even up to 10 per cent in the presence of fish;

☆ Selected varieties of fish help in controlling the weeds, insects and snail pests and also in the biological control of mosquitoes;

☆ Paddy tillering caused by the grazing of fish and fertilization by fish excreta, result in higher rice yields;

☆ Paddy bottom being highly fertile, there is usually luxuriant growth of fish-food organisms;

☆ Fish is a source of animal protein and gives an additional income to the farmer;

☆ Concurrent culture of rice and fish achieves most economical utilization of land.

The disadvantages of growing rice and fish together are quite few. These are:

1.. rice-fish culture needs elevated strong dikes, construction of trenches, canals and sumps involving additional cost; there is loss of about 5-10 per cent of the rice fields for rice to grow;

2. presence of fish in paddies restricts the use of pesticides, majority of which are harmful to fish.

Rotation of Rice and Fish

This system is known as the rotational cropping of rice and fish in paddies. It has several advantages, such as:

☆ Better cultural techniques for both can be undertaken without any detrimental effect on each other;

☆ Pesticides can be used for higher yield of rice;

☆ System permits greater water depth for fish production;

☆ Digging of trenches and canals in the paddies is not essential and saves money and space;

☆ Better control of insect pest is advised because their life cycles are disrupted;

☆ Residues from fish culture act as fertilizer for rice plants and rice stubble, when submerged, rot and encourage growth of fish food organisms.

The rotational cropping system may be subdivided into (1) single annual crop of fish in rotation with the single annual crop of rice; (2) growing fish as an intermediate crop between the harvest of the first crop of rice and planting of the second crop.

Species of Fish Used in Rice-Fish Culture

A number of fish species are being used in various countries depending on the availability and suitability for culture under local conditions. The Java tilapia and common carp are common. Nile tilapia, a better species, has been introduced recently.

Status of Rice-Fish Culture in a Few Countries

Indonesia is one of the pioneer countries where rice-fish culture is widely practised. Fish yield ranges from 23–1,202 kg/ha with an average of about 600 kg/ha/yr.

In China, peasants raise fish is over 556,000 ha of rice fields, harvesting an average of 100-150 kg/ha/ha of fish with the maximum of 700 kg/ha. They obtain quite a good profit from fish.

In India, the traditional method of culturing brackishwater fishes and shrimps in low-lying fields of West Bengal and Kerala, gives a production from 100-200 kg/ha.

In Japan, average production of carp, the dominant species cultured in rice fields, is about 100-200 kg/ha. Production, however, increases to 700-1,000 kg/ha when supplementary feeds are provided.

In the USA, fish culture in rice fields is concentrated mostly in the southern states of Arkansas, Louisiana, Mississippi and Texas where rotation of rice and fish in the same field is practised. After the harvest the fields are inundated and fish is cultured for 2 years. A production of 700 kg/ha of fish from rice fields have been reported, the average yield being 199 kg/ha/hr.

In Madagascar, the only country in Africa where traditional rice-fish culture had been practiced for many years, a yield of 200-250 kg/ha of fish is easily obtained.

Pesticides: A Constraint to Rice-Fish Culture

With the increase in the use of. pesticides in rice plantations, rice-fish culture suffered as most of these proved highly toxic to fish life. As a result, there was a sudden drop in fish production from the paddy fields. Japan suffered a major decline in her production of carps from 4,400 tons in 1943 to a meagre 462 tons in 1956. Fish yield from rice fields in several other countries also had similar fate.

In a toxicity study of various pesticides commonly used in rice fields, it is observed that most of these, especially the organochlors and organosphosphates are highly toxic to fish and a few, such as Endrin do not get degraded and leave residual effects for a long time. Recent discovery of carbo-furan, a pesticide belonging to the carbamate group, has demonstrated that it can be used without much harmful effect in fish, especially, when the mode of application is through rootzone or soil.

Economics of Rice-Fish Culture

Very little information is available on the economics of rice-fish culture system since its practice is still confined mainly to the traditional methods. It is also very, difficult to gather reliable data on fish production from rice field since a considerable portion of fish produced is consumed by the farmer and hardly any proper record is maintained. In the Philippines, some studies on the economic feasibility of rice-fish culture was carried out at Central Luzon State University (CLSU) indicating that the system is profitable. Other studies made on the technical input-output relationships in rice-fish culture also show that the system can be a profitable venture.

Research on Rice-Fish Culture

Studies on the present status of research show that very little research has been done on the improvement of the techniques of rice-fish culture.

In the Philippines, since the mid-1970's, some systematic research has been carried out at CLSU to develop low-cost appropriate technology for the production of fish in rice fields using the insect-resistant HYV's. In India, some research on rice-fish culture is carried out by CIFRI in collaboration with CRRI &.CSSRI. The result of experiments on rice and fish production from coastal saline soil areas is highly encouraging. In Indonesia, IFRI at Bogor has been engaged in rice-fish culture research since 1976.

Since its inception 25 years back the International Rice Research Institute (IRRI) has been conducting intensive research on rice culture. It developed a number of high yielding strains of rice and earned reputation for its contribution to the increase in rice production in the developing world. Notable success was also achieved by the Institute in developing insect-resistant HYV's which are mare suitable for fish culture in rice fields, and this has provided a fresh impetus to rice-fish culture.

IRRI has given considerable importance to its deep water rice breeding program in its plan for the third decade (1980's) and is developing varieties and assorted technologies to increase yields of deep water rice. If the water level in the rice field could be kept as high as up to 1 m then yield of fish could be increased considerably even within' the short culture period of 34 months. For such culture, the quicker growing, hardy species Nile tilapia is ideal and fish production could be further increased by stocking all-male Nile tilapia (monosex) and also by providing manures and supplementary feeding.

Important Areas for Further Research

The following important areas of research should also receive immediate attention of the rice-fish culturist:

1. Genetical research to evolve new strains of fish which are hardy, have high tolerance to pesticide's, arid would grow well in shallow waters;

2. Further research to find out efficient substitutes for the highly toxic insecticides;

3. Research to be carried out on the land-water use conflict.
    This needs proper engineering designs of rice fields for
    the benefit of both fish and rice cultivation; and

4. Further research on the cost benefit analysis on the rice-
    fish culture technology.

Prospect for Future Development

In conclusion, it may be said that, in spite of the constraints as
mentioned earlier, rice-fish culture is reviving with greater vigor.
The low-output captural system, as well as the simultaneous culture
system, are gradually being abandoned in favour of the more efficient
rotational cropping system of growing rice and fish, either as alternate
crops or 2 crops of one followed by one crop of the other, depending
on the condition of rice fields and availability of water. Rice-fish
culture received further imptus with the development of insect-
resistant high yielding varieties of rice coupled with the discovery
of new pesticides less toxic to fish and with improved methods of
their application. Also, introduction of the efficient, quicker-growing
Nile tilapia for rice-fish culture systems has added to the yield of
fish from rice fields.

It is hoped that with further intensive research and through
joint ventures by both rice and fish culture technologists, the present
constraints of rice-fish culture could be totally removed and
production of both commodities enhanced considerably. The
improved techniques thus developed could be extended to extensive
areas which are not utilized at present.

3. Role of Chinese Grass Carp *Ctenophyryngodon idella* (Val.) in Biological Control on Noxious Aquatic Weeds in India: A Review

Introduction

In India, according to recent estimates (Mitra, 1964; Philipose,
1968), about 320,000 ha have to be cleared of weeds annually for
fish culture out of nearly 800,000 ha of the available cultivable waters.
The problem is severe in the eastern States where 40-70 per cent of
water area is infested by weeds (Philipose *et al.*, 1962). Weed
infestations tend to convert valuable low lying inland waters into
swamps The adverse effects of excess vegetation in fishery waters
are summarized by Ling (1960).

In India, the Central Inland Fisheries Research Institute, besides other state and private organizations, has been engaged in the development of methods of weed control suitable in Indian conditions. A major step taken by the Institute in this direction was the introduction of the Chinese grass carp, which has gained popularity throughout the world as an effective biological agent for aquatic weed control.

Introduction of Grass Carp into India

Experimental consignments of grass carp from Hongkong were introduced at its Pond. Culture Sub-station, Cuttack, during December, 1959 and January, 1962 with the object of assessing its response to the new environment. The cultural qualities of grass carp, vis-a-vis indigenous carps have been carefully studied and the species has been found to be invaluable addition to indigenous culturable species (Alikunhi and Sukumaran, 1964).

Biology of Grass Carp

According to Opuszynski (1972), grass carp grows to a maximum of above 30 kg and to a length of over 100 cm. Though it is a warm water fish it could flourish in both temperate and cold climates (Hickling, 1965).

Opuszynski (1972) observed that the upper lethal temperature for grass carp fry was 38-39°C and at the same time the fish could tolerate a minimum water temperature of 1-2°C. In Arkansas, Stevenson (1965) observed grass carp could survive in a small pond under a solid ice cover for 5 weeks and in summer temperature up to 35.6°C. The lethal level of dissolved oxygen for fry of grass carp may be as low as 0.32-0.6 mg/1 under experimental conditions (Opuszynski, 1972) whereas Stevenson (1965) puts it at 0.5 ppm for adult fish. The grass carp can also withstand considerable changes in salinity and thrives well in brackish water (Gidumal, 1958). In India, Singh *et al.* (1967) observed tolerance by juveniles of farily wide fluctuations in physical and chemical factors of water.

Size and age of grass carp at first maturity varies geographically. According to Hickling (1967), in its native country (China) first maturity was reported at 4-5 years but at 9-10 years of age at Moscow. Generally, the size of the mature fish has been above 6.0 kg, but in India and Malacca grass carp matures at the age of 2 years and of small size.

Food and Feeding Habit of Grass Carp

Though grass carp is a phytophagous species, its food and feeding habits in the early stage are almost identical to the Indian major carp fry. According to Nikolsky (1956), the macrophyte feeding habit of grass carp is first observed when the fry are 17 to 18 mm. From 27 to 30 mm size the fry switch over to a predominantly vegetarian feeding habit with a negligible proportion of animal food in the diet. Singh *et al.* (1967) observed grass carp of 27 mm to 42 mm size consuming smaller duck weeds though they still relish zooplankton.

Advance fingerlings, juveniles and adult grass carp are predominantly phytophagous but their preference to different plants changes with growth. In the initial stages, it consumes filamentous algae, then small duck weeds and as the fish grows in size, it takes other leafy aquatic plants (Singh *et al.*, 1967; Opuszynski, 1972).

Though grass carp is predominantly a vegetarian fish it can also subsist on other food materials. Nikolsky (1956), quoting Chen and Lin (1935), states that grass carp reared in Chinese ponds are omnivorous and its menu may include besides various aquatic plants, land grasses, tree leaves, fruits, various oil cakes and brans, silkworm pupae etc. Under Indian conditions too it has shown similar behaviour.

The food selection depends, among other factors, on the size of the fish and its capacity to consume. According to Singh *et al.* (1967) grass carp has almost equal preference for floating weeds such as common duck weeds, *Wolffia, Lemna, Spirodela* and *Azolla*. *Eichhornia* and *Pistia* were not consumed even when pieces were fed; however, Ling (1960, 1967) reported the use of *Salvinia, Pistia* and young leaves of Eichhornia as feed for grass carp. Control of Eichhornia under experimental conditions has been observed by Avault Jr. (1965).

A tentative order of preference of the fish is Hydrilla, Najas, Ceratophyllum Ottelia, Nechamandra and *Vallisneria* (Singh, *et al.,* 1967). *Potamogeton* and *Halophyla* are also consumed by the fish. Grass carp of about 0.5 kg above consumes Nechamandra. Though the consumption of all these weeds by grass carp has been reported by other authors, opinions differ regarding the order of preference.

Among the common emergent weeds, *Trapa, Myriophyllum, Limnophila* and leaves of *Monochoria* and *Ipomoea* are also preferred by grass carp, almost in the same order. *Nymphaea, Nelumbo* and *Nymphoides* are not, however, consumed by the fish It has often been observed that in ponds where marginal weeds like *Marsilea quadrifolia, Ceratopteris thalichoides, Panicum fluitans, Cyperus* sp., *Scirpus* sp., *Jussiaea repens* etc., were present, they were cut and tender parts eaten by grass carp keeping the pond margins trimmed.

Efficacy of Grass Carp in Controlling Aquatic Plants

The important factors responsible for the effective control of aquatic plants by grass carp are water temperature, aquatic weed species composition, size and stock density of the fish (Opuszynski, 1972).

Michevicz *et al.* (1972) quoting Stroganov (1963) reported that grass carp was observed to start active feeding at water temperatures 7 to 8°C and feed intensively above 16°C. According to Hickling (1965), 22°C is the maximum limit beyond which the feeding may decline. According to him in temperate regions a faster growth rate is possible. Contrary to these observations, maximum intake of food was observed between 22-23°C whereas, below 12°C and beyond 30°C feeding ceased (Opuszynski, 1972). The performance of grass carp in clearing aquatic weeds under Indian conditions supports the findings of the latter author.

Though the weed consumption by grass carp depends on several factors, the daily intake as observed by Opuszynski (1972) amounted to 50 per cent of the weight of fish and a maximum utilization of 110-120 per cent was reached at 22-23°C and lesser consumption above this temperature. However, Singh *et al.*, (1967) observed under Indian conditions that grass carps utilized up to 286 per cent of their initial weight during a period of 41 days in July-August when the water temperature normally ranges between 27°C and 33°C.

The size of grass carp is important; fingerlings utilize duck weeds readily but may take considerable time in tackling infestations of larger plants like Hydrilla. Nechamandra, Salvinia etc. The larger fish are efficient consumers of chese aquatic weeds.

Control of Aquatic Weeds by Grass Carp

Judicious manipulation of stock density and selection of correct size of grass carps depending on the nature of water body type and quantum of weed infestation, are considered important for achieving biological control of weeds by this species.

Stock density of grass carps should be determined by a quantitative and qualitative estimate of weed density present.

In Small Water Bodies

Satisfactory results could be expected in clearing submerged weed infestation like *Hydrilla*, *Najas*, etc., when grass carps of somewhat bigger size are released, Singh *et al.* (1967) found that grass carps of 62-2,640 g size were effective against most of the common duck weeds and submerged weeds in small ponds within a short time, under stocking densities varying from 250 to 5200 fish/ha.

Kuronuma and Nakamura (1958) observed that grass carps weighing 35 kg with a stocking density of 30 fish/ha could ensure complete clearance in a pond. Aliev (1963) stated that grass carps weighing on an average 110-200 g and stocked at 30.6 kg/ha could clear 12.5 tonnes of submerged weeds in a short time. Alikunhi and Sukumaran (1964) observed that the fish could clear within a month, ponds choked with Hydrilla when they were introduced at 300-375 in numbers and 78.8 to 173 kg/ha. After they have cleared the weeds they can be transferred to some other water, leaving behind a few for maintenance of clearance. A few grass carps released along with the indigenous carps in fish culture ponds will not only keep the ponds weed free but also add to the production.

In Large Water Bodies

In large bodies of water where common predators may abound in plenty, grass carps big enough to escape predation should be stocked.

The Bhilai Steel Plant in Madhya Pradesh faced with an acute weed problem in their cooling and circulating tank with a water spread of about 2.1 km². Overgrowth of submerged weeds like Hydrilla, Najas, Ceratophyllum and Vallisneria hindered proper circulation and prevented the proper flow of water in the pipes. Futile attempts for many years were made by the authorities to clear the weeds by mechanical means. During 1971, 600 grass carps fingerlings, supplied by the C.I.F.R. Substation, Cuttack, were released on experimental basis and were found effective in checking weed growth. A further consignment of about 10,000 advanced fry were provided during 1972, for large scale introduction. According to the report, within one year the over-growth of submerged weeds in the tank was effectively checked.

The Steel Plant at Bokaro, Bihar is also facing similar weed problem in its cooling and circulation tank and has been supplied about 2,000 grass carp fingerlings for preliminary trials.

Another instance where grass carp is widely employed for large-scale control of aquatic weeds is the irrigation canal system of the Chambal Development Project, Rajasthan. The weed problem especially that of Hydrilla, has been formidable in the canals extending about 4/5 km and wetted surface of about 1165 ha. This Institute initially supplied about 600 fingerlings in July 1970 and a further 75,000 fingerlings during August-September, 1973. The impact of this species in the Chambal canal system may only be felt in the future.

Blackburn *et al.* (1971) concludes that in United States grass carp may be used to solve some of the aquatic weed problems.

In a lake of 154 ha in Poland, receiving heated waters, grass carp of 250-1000 g and numbering about 5,000 could considerably reduce the overgrowth of weeds (Opuszynski, 1972)

Performance of Grass Carp in Composite Culture vis-a-vis Utilization of Weeds for Fish Production

Grass carp is extensively used as a compatible species along with indigenous fast growing species not only in China but also in many South East Asian countries. (Lin, 1954; Hickling, 1962; Hora and Pillay, 1962).

The advantage in utulizing grass carp as compatible species in composite culture is that it occupies an ecological niche which otherwise remains unfilled by indigenous carps The herbivorous feeding habit of this species brings in the problem of feeding the fish with desired aquatic weeds from external sources if they are not naturally available in the pond to meet the food requirement of the stock. The weeds are thus converted into more useful fish flesh by the species.

Due to the low assimilation of food, the activity of grass carp results in a quick passage of digested and semi-digested plant mass into the medium thereby manuring the ponds. This accelerates the growth of phytoplankton.

The role of grass carp in enhancing fish production from fish ponds in India is encouraging. Lakshmanan *et al.* (1971) observed that grass carp representing 7.5 per cent to 17 per cent of the initial total stock density have attained an average weight of 501 to 1307 g in one year. The total weight of aquatic weeds supplied ranged from 4375-14662 kg/ha/yr. Singh *et al.* (1972) report that grass carp representing 16.67 per cent to 22.22 per cent of the initial stock density have attained an average 639 to 1250 g in one year. The quantity of aquatic weeds supplied ranged from 10,070 kg to 17,800 kg/ha/yr.

Information on food co-efficient in grass carp is scanty. The food co-efficients as observed by Opuszynski (1972) for one year old fish ranged from 36.2-35.3 at 22 to 28°C under controlled conditions. The observations of Verigin *et al.* (1963) were almost identical with those of Opuszynski. From the data presented by Singh *et al.* (1967) it could be roughly estimated that the food co-efficient of grass carp in natural environment generally ranged from 81.20 to 22.89. According to Stott and Orr (1970) the conversion rate of aquatic weeds to fish flesh by grass carp was about 224:1 which appears very low.

Propagation of Grass Carp

Availability of fish fry in adequate quantity is a prerequisite for control, especially as this species normally does not breed in confined waters. The problem of supply of fry in India has been solved by inducing the spawning by hypophysation technique (Alikunhi *et al.*, 1965; Chaudhuri *et al.* 1966; 1967 and Singh *et al.*, 1970). Successful rearing of fry and fingerlings to produce stock material has been achieved now (Singh, *et al.*, 1972).

Suggestions for Speedy Distribution

Though the fry of grass carp has been supplied to different state departments and other agencies since 1964 for its propagation in different parts of the country, little progress has been achieved in this direction. Very few state departments could so far succeed in breeding and building up their own stock in spite of the intensive training imparted to fishery department personnel and publications on the subject by the Cuttack Sub-station of C.I.F.R. Institute.

The progress of large scale production and utilization of grass carp for biological control on an intensive scale depends mainly on the concerted efforts by the various state departments and other agencies interested in this species. This problem should be tackled on a "war footing". Each state should develop a few centres manned by sufficiently trained personnel in breeding and rearing the species.

Summary

The Chinese grass carp introduced in India, in recent years, is gaining popularity not only as an important species in composite fish farming but also as an efficient biological agent for controlling weeds. In water supply tanks of neavy industries where the use of chemical weedicides is prohibited, eradication of submerged weeds by grass carps is the most effective and cheap control method.

The efficacy of grass carp in controlling aquatic weeds depends on several factors such as water temperature, species composition of weeds and the size and stock density of grass carp.

The technique of hypophysation adopted for inducing grass carp to spawn is almost the same as that adopted for indigenous carps. Rearing of fry and fingerlings and maintenance of broodstock of the species are also easy. The progress of large-scale production and utilization of this species for biological weed control on an intensive scale depends mainly on the concerted efforts by the various state fisheries departments.

Selected References

Alikunhi, K.H. and Sukumaran, K.K., 1964. Preliminary observations on Chinese carps in India. *Proc. Indian Acad. Sci.* (B), 60(3): 171-188.

Alikunhi, K.H. and Sukumaran, K.K., 1965. Observations on growth maturity and breeding of induced-bred pond-reared silver carp, *Hypophthalmichthys molitrix* and grass carp, *Ctenopharyngodon idellus* in India during July, 1962 to August, 1963. *Full. Cent. Inst. Fish. Educ. Bombay,* (2): 19.

Chaudhuri, H., Singh, S.B. and Sukumaran, K.K., 1966. Experiments on large-scale production of fish seed of the Chinese grass carp, *Ctenopharyngodon idellus* (C and W) and the silver carp, *Hypophthalmichthys molitrix* (C and V) by induced breeding in ponds in India. *Proc. Indian Acad. Sci.,* 63(B)(2): 80-95.

Chaudhuri, H. *et al.*, 1967. Note on natural spawning of grass carp and silver carp in induced breeding experiments. *Sci. Cult.*, 33(2): 493-494.

Hickling, C.F., 1962. *Fish Culture*. Faber and Faber, London. 295p.

Hickling, C.F., 1965. Biological control of aquatic vegetation. *PANS*, 11: 237–244.

Hickling, C.F., 1966. On the feeding process of the white amur *Ctenopharyngodon idellus. Proc. Zool. Soc. Lond.*, p. 408-419.

Hickling, C.F., 1967. On the biology of herbivorous fish, the white amur *Ctenopharyngodon idella* (Val.). *Proc. R. Soc. Edinb.*, 70: 62-81

Kuronuma, K. and Nakamura, K., 1958). Weed control in farm pond and experiment by stocking grass carp. *Proc. Indo-Pacific, Fis. Coun.*, 7(2): 35-42.

Singh, S.B. *et al.*, 1967. Observations on the efficacy of grass carp in controlling and utilizing aquatic weeds in India. *Proc. IPFC*, 12(2): 220-234 (1966).

Singh, S. B. *et al.*, 1972. Observations on composite culture of exotic carps. *J. Inland Fish. Soc. India*, 4: 38-50.

Chapter 12

Compilation of SEAFDEC Aquaculture Technical Papers

A RESUME

Since 1976, research of the Aquaculture Department of the Southeast Asian Fisheries Development Centre has been centered around four major goals:

1. The development of the technique to induce maturation and spawning in wild and captive adult milk fish;

2. The development of the technique for the mass production of milkfish fry.

3. The development and improvement of existing techniques to catch, transport and store milkfish fry with minimum mortality; and

4. The development and/or improvement of existing techniques to maximize production of marketable-sized milkfish in brackishwater ponds and pens with minimum inputs.

With partial financial support from the International Development Research Centre of Canada, some of these objectives have been realized and are described in Volumes-I and II of their report.

The SEAFDEC work on Milkfish has been done with success and the details have been recorded in the author's edition in "Aquaculture 2000 — A New Horizon" works on carp in the Philippines, preservation on pituitary extracts and aquatic weeds have been distributed in this chapter.

The milkfish, *Chanos chanos* (Forsskal), is an important food fish in Indonesia, Taiwan and the Philippines. It is widely distributed in the tropical and subtropical areas of the Pacific and the Indian Oceans. It is a hardy, fast-growing euryhaline species and has a very short food chain and an efficient food conversion. Although this species has been cultured for centuries in brackishwater ponds, fish farmers are still entirely dependent on the natural fry fishery. fry are collected along with coastline during the spawning season and the annual supply of fry for stocking in ponds is unstable due to seasonal environmental conditions and natural fluctuations in yearly recruitment. Because of its importance as food and the potential of milkfish fingerlings as baits for tuna and skipjack fisheries, there has been a growing interest in breeding milkfish in captivity not only in the Philippines, Indonesia and Taiwan, but also in Hawaii, Kiribati and Tahiti.

Since 1976, research of the Aquaculture Department of the Southeast Asian Fisheries Development Centre has been centered around four major goals:

1. The development of the technique to induce maturation and spawning in wild and captive adult milk fish;

2. The development of the technique for the mass production of milkfish fry.

3. The development and improvement of existing techniques to catch, transport and store milkfish fry with minimum mortality; and

4. The development and/or improvement of existing techniques to maximize production of marketable-sized milkfish in brackishwater ponds and pens with minimum inputs.

With partial financial support from the International Development Research Centre of Canada, some of these objectives have been realized and are described in Volumes-I and II of their report.

1.   Carp Management in the Philippines

Carp culture is an age-old practice in many countries in Asia particularly in China and India. These important cultivated freshwater species constitute the major food fishes of millions of people in these countries where population explosion has largely contributed to the deficiency in protein in their diet.

Carps are preferred as an important commodity for aquaculture because of their desirable qualities such as, faster growth rate, herbivorous food habits having a short food chain and their general acceptability by the consumer public. Polyculture of a number of compatible carp species of different feeding habits, each utilizing different ecological niches in the pond, is an important management technique to obtain high production. Amazingly high yield of 10 tons ha yr and over have been reported from India and China in recent years through polyculture (Chaudhuri *et al.*, 1975; Tapiador *et al.*, 1977)

In the Philippine freshwater fish culture is at its infancy although the potential for development of freshwater aquaculture is quite high. Unlike other countries of the Southeast Asian region, the Philippines does not have a long history of carp culture. However, attempts to initiate carp culture in the Philippines have been made during the last two decades or so when a number of commercial carp species were introduced into the country.

Status of Carp Culture and Carp Seed Production in the Philippines

The common carp *(Cyprinus carpio)* was the first carp species introduced and successfully propagated in the Philippines. It has established itself in the lakes, rivers and other freshwater impoundments. Since early sixties several Asiatic carps have been introduced into the country. These are the Chinese grass carp *(Ctenopharyngodon idella)*, silver carp *(Hypophthalmichthys moliffix)* and bighead *(Aristichthys nobilis)* and the Indian carps catla *(Catla catla)*, rohu *(Labeo rohita)* and mrigal *(Cirrhinus mrigala)*. Of the Indian carps rohu alone has established in the lake Laguna de Bay and also introduced in several regions where it is reported to have bred. The Chinese carps are also reported to have established in the river Pampanga. More recently, the crucian carp *(Carassius carrassius)* has been transplanted in Pampanga river from Japan where the species has established itself.

To assure adequate supply of seed of these imported carps the Government as well as private fish hatcheries were established where induced breeding of Asiatic carps were conducted. The bighead and silver carp were successfully bred in 1969 and the grass carp was induced to breed in 1971 (Reyes, 1975). More recently, the Tanay Fisheries Station of BFAR in Rizal is being engaged in the hypophysation of silver carp, rohu and bighead. Experimental studies have been undertaken at the Freshwater Fisheries Station of the SEAFDEC Aquaculture Department at Binangonan/Tapao Point on the common carp, rohu and silver carp, particularly on their stocking rate and growth in cages and pens in the lake environment. The new hatchery under construction at the station is primarily designed for the mass production of seed of the Asiatic carps for making seed available for culture in the freshwaters of the country. Extensive studies have also been undertaken at the Freshwater Experimental Fish Farm of the Central Luzon State University (CLSU) on the breeding" and performance of these introduced species in Philippine waters.

Induced Breeding and Carp Hatchery Management

Indian and Chinese Carps

The Indian and Chinese carps do not ordinarily spawn in ponds although they mature in that environment. They naturally breed in rivers when these are in flooded conditions during the rainy season. However, under artificial conditions they are induced to breed by injections of extraneous hormones. Carp pituitary hormones are found to be very effective in inducing spawning of these Asiatic carps. However, pituitary hormones of fishes other than carps, SG-G100, HCG and mammalian hypophyseal hormones (Synahorin etc) are also successfully administered in inducing spawning of these carps. The latter hormones are very often used in combination with fish pituitary hormones. More recently, a synthetic hormone, a nonapeptide LH-RH analogue has been successfully used in breeding Chinese carps (Anon., 1977).

Induced Breeding Techniques

The Asiatic carp are seasonal breeders and spawn mainly during the rainy season. Mature spawners are raised in ponds. The sexes can be distinguished easily by feeling the inner side of the pectoral fin. It is rough in case of males and smooth in females (Chaudhuri, 1959). Pipe, freely oozing males and fully mature females with distended abdomen are selected for spawning. The females, especially of grass carp, are cathetered and a few eggs examined for maturity stage. The female spawners are given a preliminary low dose of carp pituitary extract, followed by a final higher dose at an interval of 6 hours (Chaudhuri, 1969). The dose is determined according to the weight and stage of maturity of spawners. The males are administered only one low dose at the time of second injection to the female. The sexes are next put together either inside breeding hapa fixed in ponds or in breeding tanks/circular cemented spawning pools with continuous circulation of water and/or with aeration. Within 4-6 hours the fishes are ready to spawn. In case of Indian carps, spawning naturally occurs but the Chinese carps generally require artificial fertilization. When the female is ready for ovulation, the eggs are stripped and artificially fertilized by the milt stripped from males. Usually the 'dry method' is followed (Chaudhuri *et al.*, 1966). The fertilized eggs absorb water and swell up. The water-hardened eggs are then transferred to hatchery. The Asiatic carps usually spawn at water temperature range of 24°-31°C, the optimum temperature being around 27°C.

Problems in Successful Induced Breeding

The main problems encountered in successfully carrying out induced breeding program be categorized as follows:

Non-availability of Adequate Numbers of Suitable Spawners

Failure in induced breeding experiments is very often due to poor condition of spawners selected for breeding. It is essential that adequate numbers of spawners are maintained in the fish farm and supplied with right type of food for proper development of gonads. Overfeeding sometimes lead to accumulation of extra fat in the abdomen which is responsible for ill-formed gonads in fishes.

Unfavorable Climatic and Physico-chemical Conditions of the Environment

Unfavorable weather condition is an important limiting factor in large-scale seed production. It is well-known that favorable climatic conditions, rain and optimum water temperature are conducive to successful spawning. Prolonged drought and failure of rain fail to bring the gonads of carps to a prime condition. Spawning may sometime occur at water temperature above the optimum range but the percentage of fertilization and hatching is generally poor. Indoor breeding under controlled water temperature with facilities for continuous circulation of water and aeration may help solve this problem.

Dearth of Pituitary Gland

Shortage of pituitary glands and high cost of SG-G100 and other mammalian hormones very often restrict induced breeding program. It is essential that carp breeding stations should be self-sufficient in their requirements of hormones. To solve this problem, it is suggested that the required number of donor fish may be raised in the fish farm itself. The common carp is an ideal donor fish. Since the common carp is found mature almost throughout the year, pituitary glands could be collected year round. The cost of pituitary glands is normal because the fishes could be sold after collection of the gland.

Determination of Proper Ripeness in Female Spawners

Whereas the male spawners can easily be selected by this freely oozing wilt it is sometimes difficult to select ripe females, especially in case of grass carp. In such cases,it is advisable to use catheter for collecting a few eggs through the genital opening and examine the stage of maturity of the eggs before selecting the female for hormone injection.

Insufficient Milt for Artificial Fertilization

Very often the problem of insufficient milt from males, especially of silver carp, is encountered during artificial fertilization of eggs. Silver carp males in tropical waters usually mature in one year. Such males produce very little milt. To solve the problem of insufficient milt, two to three-years old healthy males should be selected and the ratio of male and female increased to three or four males for each female.

Low Fertilization of Eggs

Failure in fertilization or low rate of fertilization may be due to faulty method of stripping. The ovulating female should be selected for stripping at the right moment when the eggs are about to ooze. Stripping should not be done by force. During stripping sufficient care should be taken to wipe the body of the spawners with a dry towel so that slime, fecal matters and water do not get mixed with the eggs and milt. The fertilization should be done immediately after collection of a batch of eggs instead of waiting for collection of all the eggs from the female.

Hatchery Management

An estimation of the quantity and percentage of fertilized eggs are made before the water-hardened eggs are transferred to the hatchery. Different types of carp hatcheries are in vogue depending on the financial capability of the fish breeder. The outdoor hatcheries either comprising of a series of double hatching hapa (Chaudhuri, 1960) or of circular cemented hatching pools with facilities for constant circulation of water (Tapiador *et al.*, 1977) are generally used. The indoor hatchery is usually provided with hatching jars or hatching funnels under controlled temperature and with continuous circulation of water and aeration. The incubation period of Indian carp eggs is about 15-18 hours at water temperature 27-31°C. The Chinese carp eggs take little longer time and the incubation period may prolong to 3648 hours at a slightly lower temperature.

Outdoor Hatchery

The fishfarmers in India usually use an outdoor hatchery consisting of a series of double hatching hapa fixed to bamboo poles in ponds. The inner hatching hapa is made of mosquito net cloth and the outer one is made of close-meshed linen cloth. The developing eggs are evenly distributed in the inner hapa. When the eggs hatch out, the hatchlings pass through the mesh of the inner hapa and escape to the outer one. The inner hapa containing the egg shells and bad eggs is removed when all the eggs have hatched out. The hatchlings are left in the outer hapa for two more days till the yolk is absorbed and the fry are ready for transfer to the nurseries.

In China, the outdoor hatchery consists of circular cemented hatching pools. These are either single or multichambered and are provided with diagonally-installed pipes at the bottom of the pools for constant circulation of water. Portable hatching jars are also used for incubation.

Indoor Hatchery

The indoor carp hatchery used in the fish breeding stations in India, and in a few countries usually consists of a series of hatching jars (made of glass, plastic or fiberglass) provided with continuous flow of water. The jars are cylindrical above ending in a cone-shape at the bottom. They are open at the top and having a small hole with a nozzle at the bottom through which a rubber tubing is fitted. Water enters through the rubber tubing into the jar and flows out through a beak fitted at the top of the jar. The developing eggs are introduced into the jar. The flow of water is so regulated that the eggs are not carried away out of the jar but are constantly in circulation inside the lower two-thirds of the jar. The hatchlings, however, have the habit of vertical movement and as soon as they come near the surface they pass out along with the flow of water and collect in the spawnery. A spawnery is a hapa fitted in a tank with continuous flow of water and aeration. Egg shells, bad eggs and deformed hatchlings are left behind in the jars and are removed by disconnecting the jar from the top. The number of eggs introduced in each jar depends on the capacity of the jar which may vary. A 13 liter capacity jar may hold 100,000 eggs at a time. The hatchlings are left in the spawnery for two days till the yolk is absorbed and the fry are ready for stocking into nurseries.

In some carp hatcheries instead of jars, seive-cloth funnels or plastic funnels are used.

Problems in Carp Hatchery Management

1. In outdoor hatchery especially those without aeration or circulation of water, mortality during incubation occurs due to sudden rise or fluctuations in water temperature, presence of algal blooms producing oxygen bubbles during photosynthesis resulting in buoying up of eggs to the surface, poor oxygenation inside the hapa, direct destruction of eggs by certain fishes present in the pond and also destruction of hapa due to storms and typhoons.

2. Source of water is very important for indoor hatchery. Deep tube well water is usually low in oxygen and high in free $CC^\wedge$ content. This water has to be run through open channel before supplying to the hatchery for oxygenation and getting rid of high CO_2 content which is harmful for eggs. If the water source is a river then the water need to be filtered for eliminating silt as well as eggs and fry of miscellaneous fishes which are harmful for the carp hatchlings. The water pumped from the ponds or reservoir also has to be sieved properly or treated with UV to prevent and kill the copepods (cyclopidae) which are highly destructive to carp hatchlings.

3. Soft water is usually responsible for breaking up of egg shell resulting in premature hatching of eggs. Besides, very often the developing eggs-get infected with bacteria and fungi and in most cases premature hatching occurs resulting in very poor survival of hatchlings. In such cases, the developing eggs may be treated with a solution of 5-10 g tannin in 10 liter clean drinking water (Woynarovich, 1975).

4. Destruction of hatchlings due to fungus infestation: The bad eggs and eggshells, if not separated from the hatchlings, the latter is liable to get fungal infected. Large-scale mortality of hatchlings occurs due to such fungal infections.

Common Carp and Crucian Carp

The common carp matures in the first year and breeds almost throughout the year with the peak breeding seasons in spring season (March-April) and rainy season (July-August). They breed in ponds and attach their eggs to grasses, aquatic weeds or any other substrata. Survival of young ones due to such wild spawning is very low. Hence special breeding technique is followed for obtaining higher survival. The breeding habit of crucian carp is more or less similar to that of the common carp.

Breeding and Hatching Technique

To prevent wild spawning, it is essential to segregate healthy male and female adult fish and raise them to spawners in separate ponds. Fully ripe healthy spawners are selected for breeding. Two to four males are usually put along with each female and introduced into a breeding hapa or a breeding tank freshly filled with clear water. The breeding container is supplied with suitable materials for attachment of small sticky eggs of common carp. The materials may either be some submerged aquatic weeds such as, *Hydrilla* or *Najas* or an artificial egg collector called *Kakabans*. Within 10-12 hours spawning is usually completed. The fertilized eggs attached to the egg collectors are then transferred to hatching hapa or hatching tanks for ensuring higher survival of hatchlings. Hatching tanks are provided with aeration device maintaining a flow of water. The incubation period is about 48 hours at water temperature 27°C-31°C and may prolong to 72 hours at lower temperature.

Problems in Breeding and Hatchery Management

Problems in breeding common and crucian carp and in their hatchery management are basically same as those for the Asiatic carps. Maintenance of an adequate stock of spawners is very important. Clear oxygenated water is essential for both breeding and hatching environment.

Nursery Management

The three-days old (5-7 mm) post-larvae of carps are transferred from the hatchery to nursery ponds for rearing. Where there are facilities for indoor rearing of larvae in the hatcheries, the postlarvae may be reared initially for 5-7 days in troughs, tanks or raceways with proper aeration and fed heavily with zoo-plankters such as rotifers and cladocerans till they grow to 12-15 mm when the fry may be transferred to prepared nursery ponds.

Nursery rearing of post-larvae (fry) is one of the most important phases of carp culture. Improper management at this phase may lead to high mortality resulting in very poor survival. Proper attention is essential at every operational stage of nursery management such as 1) pond preparation 2) stocking and post-stocking care, and 3) harvesting. Nursery management is basically same for the Asiatic carps as well as the common carp and crucian carp.

Pond Preparation

Nursery ponds are usually very small-sized shallow ponds of 0.02 to 0.05 ha in area with water depth of 0.9 to 1.2 m. It is easier to manage small size seasonal ponds. The following series of operations are necessary for proper preparation of nursery ponds.

Weed Clearance

Nursery ponds are normally cleared of vegetations by manual labors. Ponds should be kept free of weeds till the fry are harvested.

Eradication of Unwanted Fishes and Other Organisms

The main cause of fry mortality in nurseries is due to predation caused by miscellaneous fishes present in the pond and other animal life. These enemies are eradicated either by draining or by the application of a fish poison or by both. Usually a fish poison of plant origin is recommended such as, tobacco dust, tea-seed cake (Tang, 1961), derris root powder, mohua oil cake (Chaudhuri, 1971) etc. These are mixed with pond water and the fishes are killed within a few hours. Fishes killed by piscicide of vegetable origin are fit for human consumption. Any chemical pesticides should not be used because they are both harmful to human beings as well as to the fertility of the pond soil and toxicity continues for a long period.

Liming and Pond Fertilization

Liming of nursery pond is necessary for increasing the pH of acidic soil and also for disinfecting the pond. This is followed by pond fertilization usually by organic manures such as cow dung, poultry manure, pig wastes etc. It is essential to maintain a continuous bloom of zooplankters in nursery ponds which are the main food items of carp fry. In case of depletion of zoo-plankters inorganic fertilizers such as, urea and superphosphate may be intermittently added. The dose of manure depends on the fertility of the pond soil.

Eradication of Aquatic Insects

Aquatic insects especially the backswimmers are present in large numbers in manured nursery ponds. These are highly predatory and can cause immense damage to the 5-7 mm size three-days old fry. Majority of larger insects can be removed by slow drag-netting with a fine-meshed net. The remaining insects are killed by application of an emulsion of vegetable soil and cheap soap (56 kg: 18/ha) to form a thin film of oil on the water surface. The backswimmers are prevented to respire and are killed with 15 to 20 minutes. Oil emulsion is applied a few hours before the fry are stocked. Care is taken to apply oil emulsion at a time when the day is not windy or rainy.

Stocking of Fry and Post-stocking Care

Stocking

Stocking of fry is done a few hours after application of oil emulsion and assessment of zoo-plankters which are the main food of fry. While stocking fry, care is taken to gradually equalize the temperature of water inside the fry container with that of the pond. The stocking rate is ordinarily 1-2 million fry per hectare. The stocking rate may be increased up to 10 million per hectare depending on the food available and pond conditions.

Supplementary Feeding of Fry

Supplementary feeding is done from the second day of stocking when the zooplankters are likely to be depleted and also when the pond is heavily stocked. Usually finely powdered and sieved soyabean meals or peanut oil cake along with fine quality rice bran or wheat short is fed to the fry twice daily. The feed ration is increased at intervals of every five days. Better results are obtained when the feeds are mixed with growth promoting substance like cobalt chloride (Sen and Chatterjee, 1976).

Harvesting

Normally 5-7 mm postlarvae of carps should reach 25-30 mm in about two weeks with survival ranging from 50-80 per cent. The stock should be thinned out or totally harvested after two weeks or so, otherwise overcrowding for a long time may result slow growth rate with the attendant danger of unhealthy conditions leading to mortality due to infection and diseases. The harvested fry are then stocked in rearing ponds which is the second phase of culture and grow them there to fingerling size.

Problems in Nursery Management

In nursery management, problems may arise at every step and unless immediate proper attention is given may lead to very poor results. The main problems are the following:

Adverse Physico-Chemical Conditions of Water

While transporting postlarvae from the hatchery or fry collection centres to the nursery site and stocking them into nurseries care should be taken to determine the dissolve O_2 content of water, pH, free CO_2 and the water temperature and take suitable measures since hydrobiological conditions of ponds have a direct relationship to the survival and growth of fry.

Destruction of Fry by Predatory Fishes

Predation by miscellaneous fishes and other enemies of postlarvae and fry appears to be one of the major factors responsible for poor survival of fry in nursery ponds. Use of a suitable piscicide for eradication of the predatory and weed fishes present in the nursery pond is essential. The main problem for the fish farmers is procurement of a suitable piscicide since commercial derris-root powder (Rotenone) is not locally available. It is essential that a piscicide of vegetable origin may be made available to the fish farmers. Plants like *Croton tiglium, Millettia piscidia, Barringtonia acutangula* (Chakraborty, *et al.,* 1972), *Derris* sp. available in the Philippines may be tapped for commercial production of the piscicide. In absence of a suitable piscicide fish farmers are using highly toxic pesticides like Endrin and other chlorinated hydrocarbon products which are extremely harmful.

Predation by Backswimmers and Other Insects

Eradication of insects from nursery ponds is a major problem. Nursery ponds can be treated with any insecticide for eradicating all insects but the toxicity continues for sometime and fry cannot be stocked immediately. In the meantime the ponds get repopulated by insects. Therefore, for eradiation of insect such materials should be used which is harmless to fry so that the fry could be stocked immediately. Soap oil emulsion is the best for this purpose.

Lack of Proper Food of Fry

Mortality occurs at the early stage of nursery rearing of fry if adequate quantity of right type of food (zooplankton) is not present. Artificial feeds are useful later on but initially the post-larvae supplied with rotifers and cladocerans give much higher survival.

Faulty Post-stocking Management

While the fry are being grown, the nursery ponds may get re-infested with weeds, filamentous algae and dense algal bloom. These are to be kept under control. The inlets and outlets should be properly screened to prevent entry of miscellaneous fishes and also measures taken to prevent entry of *Channa striatus* (dalag) from adjacent waters. Harvesting of fry at the proper time is also an important post-stocking measure.

Conclusion

The present status of carp culture in the Philippines has been briefly mentioned. The technology of carp seed production including management of hatcheries and nurseries are available, the problems confronting them have been discussed, and possible solutions of the problems suggested. These observations indicate that the freshwater environment in the Philippines is quite suitable for the propagation and culture of the introduced Asiatic carps.

Carps, however, are not yet very popular as food fish in the Philippines. In the neighbouring countries of Southeast Asia, the whole of South Asia, China and Japan carps are highly esteemed food fishes. They are also very popular in many European countries, USSR and in Israel. It is expected that carps will gradually meet the consumers' preference in the Philippines also. With the increase in population the demand for fish protein will be considerably increased. To meet the demand, low-priced fishes capable of giving high yields such as carp and tilapia will have to be intensively cultured. The present average production of milkfish is roughly about 680 kg/ha/yr. With little effort an average production of 5 to 6 tons is attainable through polyculture of carps. Preliminary experiments conducted in the Philippines on polyculture of three species of carps (common carp, rohu and silver carp) gave satisfactory results (Grover and Banacia, 1973). It has also been reported that production of over 3 tons/ha/yr was obtained in an upland farm in the Philippines through culture of Chinese carps (Reyes, 1975). These results are highly encouraging.

Carp culture can easily be integrated in the rural development program of a developing country. Polyculture of carps has been successfully demonstrated as an efficient waste treatment system. Whereas in intensive aquaculture, fertilizers and feeds are the major constraints, use of animal wastes as a substitute for fertilizer in carp ponds for the production of natural feed is likely to contribute greater interest in these species.

(Source: H. Chaudhuri in SEAFDEC Publication, 1977)

Selected References

Anon., 1977. A new highly effective ovulating agent for fish production–practical application of LH-RH analogue for the induction of spawning of farm fishes by Cooperative team of hormonal application in pisciculture. *Scientia sinica*, 20(4): 469-474.

Chakraborty, D.P., Nandy, A.C. and Philipose, M.I., 1972. *Barringtonia acutaagula* (L.) Gaertn. as a fish poison. *Ind. J. Expt. Biol.*, 10(1): 78-80.

Chaudhuri, H., 1959. Notes on external characters distinguishing sex of breeders of the common Indian carps. *Sci. and Cult.*, 25(10): 258-259.

Chaudhuri, H., 1960. Experiments on induced spawning of Indian carps with pituitary injections. *Indian J. Fish.*, 7: 2048.

Chaudhuri, H., 1969. Techniques of hypophysation of Indian carps. FAO/UNDF Regional Seminar on Induced Breeding of Cultivated Fishes, FRI/IBCF/11: 5 p.

Chaudhuri, H., 1971. Stocking pond management in composite fish culture. In: *All-India Coordinated Research Project Workshop on "Composite Fish culture"*, Cuttack (mimeographed): 14 p.

Chaudhuri, H., Singh, S.B. and Sukumaran, K.K., 1966. Experiments on large-scale production ot fish seed of the Chinese grass carp *Ctenopharyngodon idellus* (C. and V.) and the silver cap

Chaudhuri, H., Chakraborty, R.D., Sen, P.R., Rao, N.G.S. and Jena, S., 1975. A new high in fish production in India with record yields by composite fish culture in freshwater ponds. *Aquaculture*, 6: 343-355.

Sen, P.R. and Chatterjee, D.K., 1976. Enhancing production of Indian major carp fry and fingerlings by use of growth promoting substances. Contributed to FAO Technical Conference on Aquaculture, Kyoto, Japan, 26 May-2 June 1976. FIR: AQ/Conf/ 76/E, 66: 120.

2.   Preservation of Pituitary Extract

The concept of "Pituitary Hormone Bank" and conservation of it's extract struck Prof. Chaudhuri immediately after his success of induced breeding in 1957.

The editor presents one of his papers published jointly by K. H. Ibrahim and H. Chaudhuri as early as in 1966. Prof. Chaudhuri also worked on hormone bank while in SEAFDEC.

Preservation of Fish Pituitary Extract in Glycerine for Induced Breeding of Fish

Fish pituitary extracts (20-40 mg. pituitary gland/ ml.) of major carps, *rohu, mrigal* and *catla*, prepared in distilled water-glycerol (1:2), retain their potency even after storage for periods varying from 9 to 61 days at room temperature or under refrigeration. Intramuscular injections with pituitary extracts of *mrigal* and *catla* induce 100 per cent spawning and of *rohu* 52-6 per cent; spawning is obtained in *mrigal* with homoplastic gland extracts and in *rohu* with both homoplastic and heteroplastic gland extracts. Extracts prepared prior to the commencement of the breeding season can conveniently be used during the entire breeding season. Ampouling the extract is a better mode of preservation for commercial production.

Considerable progress has been made in India, particularly in the latter half of the last decade, on the technique of administering fish pituitary hormones for successful spawning of Indian and Chinese carps for cultural purpose. In view of the high admixture of undesirable species very often observed in carp spawn collections from riverine sources and owing to uncertainties in these collections, it is recognized that induced spawning of carps would become the major source of fish seed production in the country. To achieve this, besides perfecting the technique and making it quite dependable, the method employed has to be simplified so that the fish farmers themselves could successfully adopt them.

Extracts prepared from alcohol preserved pituitary glands in distilled water or 0-3 per cent saline at the time of injection are administered immediately. Such extracts, when once prepared, cannot be kept for long. Preliminary attempts at preservation of pituitary extracts without affecting their effective potency and their application in induced breeding of carps are reported in this communication.

A known quantity of pituitary glands was macerated in a tissue homogenizor. The volume of extract for the weight of glands was calculated at a fixed concentration and distilled water equal to a third of this total volume was added to the macerated glands and thoroughly mixed. The solution was kept in a refrigerator for about 24 hr after which pure glycerine was added to make up the volume, maintaining a 1: 2 ratio of distilled water and glycerine. The suspension kept in the refrigerator for another 24 hr was taken out and centrifuged. The supernatant liquid, filled in airtight bottles or sealed in ampoules, was kept under refrigeration or at room temperature for later use. Five batches of extracts were thus prepared with concentrations of 20 to 40 mg./ml. from pituitary glands of major carps (*rohu* 3; *mrigal* 1; and *catla* 1) preserved in absolute alcohol for 23-78 days.

Using the above extracts 34 sets of experiments were conducted during July-August, at Killa Fish Farm, Cuttack, on induced bred and pond reared breeders of *rohu* (*Labeo rohita*) and *mrigal* (*Cinhina mrigala)*. Water temperature during the experimental period varied from 27-5° to 34-6°C.

The female received a preliminary dose of 2-3 mg./kg. body weight and a second dose of 6-11 mg./kg. body weight after 6 hr. In some instances a third injection, 12 hr after the second, was given to induce spawning and in such cases the dose was either similar to the second or slightly higher. Males were given only a single dose of 2-3 mg./kg. body weight at the time the females received the second dose. The extracts were administered intramuscularly.

Successful spawning was obtained in *mrigal* with injections of homoplastic glands and in *rohu* with both homoplastic and heteroplastic glands. Spawning took place usually 4-6 hr after the second injection or 3-5 hr after the third. These fishes reacted similar to those receiving normal pituitary injections, and there was no visible adverse effect on the injected breeders till the completion of the experiments, although Clemens and Sneed have mentioned about the physiological effects of administering hypertonic glycerine solutions on fishes. The number of sets of breeders injected and results obtained are given in Table 1. It is interesting to note that the extracts from four batches induced spawning in 25 sets injected after different intervals of preservation. But fishes injected with extract from one batch did not–respond. These females on autopsy showed a fully ripe ovary and there was no visible effect of the hormones. No probable explanation could be put forth for this observation.

Table 1: Effect of Fish Pituitary Extract on Spawning

Pituitary Extract from	Weight of Recipient Females (kg.)	Total Number Injected	Positive Results (%)	Extract Preserved (Days)
Mrigal	0.75–2.25	13	100.0	11–36
Catla	1.0–3.0	2	100.0	16
Rohu	1.0–2.5	19*	52.6	9–61

*: This includes 9 sets injected from one batch of extract and no spawning was obtained.

As distilled water-glycerine extracts of the pituitary glands maintained their potency when stored in airtight vials, ampouling of the extracts was tried. *Rohu* glands (362-6 mg.) were extracted in 9 ml. of solution and the extract was sealed in 9 ampoules with 40 mg./ml. per ampoule and stored for 11 days. Four sets of *rohu* were injected with extract from these ampoules and all the four sets spawned successfully yielding fertilized eggs. Ampoules stored both at room temperature and in a refrigerator yielded successful results. Thus ampouling the extract makes preservation and storage more efficient and convenient, and commercial production easier.

Preservation of pituitary glands in absolute alcohol and extraction of hormones in distilled water or in 0-3 per cent saline at the time of injection are routine practices adopted in India for induced breeding of carps. Ribeiro and Neto and Neto and Ribeiro have reported preparation of fish pituitary extract in glycerine. Neto and Ribeiro found that when such extracts were kept in sealed airtight vials at room temperature, they retained their potency for a year. De Menezes *et al.* observed the extracts of gonadotrophins in glycerine more uniform in potency than those in saline or distilled water.

The present experiments have shown that carp pituitary extracts in glycerine can be conveniently prepared in advance of the breeding season and used as and when needed. For example, an extract prepared in the month of June can be conveniently and effectively used during the entire breeding season.

The method described is much simpler and easily adaptable by fish farmers and has the following distinct advantages over the existing procedures: (i) The time required for selecting and extracting the required weight of glands for each injection is saved; (ii) wastage of glands/extract is avoided; (iii) extraction of large number of glands at a time ensures uniform potency of hormones for unit volume of extract, eliminating possible differences in potency of glands from donors at different stages of maturity as well as due to the imperfect preservation of donor fishes. This ensures better and more dependable response to injections of the extract; and (iv) enables a larger number of fish being handled in a given time, making commercial fish seed production easier.

(Source: K. H. Ibrahim and H. Chaudhuri,
Indian Journal Experimental Biology)

Selected References

Alikunhi, K.H., Sukumaran, K.K. and Parameswaran, S., 1962. *Proc. Indo-Pacif. Fish. Coun.*, 10th Session, Sec. II, 181.

Alikunhi, K.H., Sukumaran, K.K., Parameswaran, S. and Banerjee, S.C., 1964. *Bull. cent. Inland Fish. Res. Inst. India*, p. 3.

Alikunhi, K.H., Vijayalakshmanan, M.A. and Ibrahim, K.H., 1960. *Indian J. Fish.*, 7(1): 4.

Chaudhuri, H. and Alikunhi, K. H., 1957. *Curr. Set.*, 26(1): 12.

Chaudhuri, H., 1963. *Proc. nat. Insl. Sci. India*, 29B: 4.

Chaudhuri, H., Bhowmick, R.M., Pal, R.N. and Munshi, A.B., 1962. *Proc. Indian Sci. Congr. (Abstr.)*, Part III, 391.

Chaudhurii, H., 1960. *Indian J. Fish.*, 7: 1.

Clemens, H.P. and Sneed, K.E., 1962. *U.S. Department of the Interior, Fish and Wild Life Service, Bureau of Sport, Fisheries and Wild Life Research Report*, p. 61.

De Menezes, R.S., Fontenele, O. and Camacho, E.C., 1957. In: *The Physiology of the Pituitary Gland of Fishes* by G.E. Pickford and J.W. Atz (New York Zoological Society, New York).

Pickford, G.E. and Atz, J.W., 1957. *The Physiology of the Pituitary Gland of Fishes* (New York Zoological Society, New York).

Ribeiro, O.F. and Neto, J.F.T., 1957. In: *The Physiology of the Pituitary Gland of Fishes* by G.E. Pickford and J.W. Atz (New York Zoological Society, New York).

3. Aquatic Weeds–Control thereof

Prof. Chaudhuri conducted a lot of experiments on the control of weeds which jeopardise the growth of fish in aquatic substratum. Aquatic weeds use up nutrients which need control.

Prof. Chaudhuri applied his aquired experience to solve a critical problem of weed devasation in Fiji and elsewhere.

The following gives a short account on the "Control of aquatic weed by moth larvae".

Control of Aquatic Weed by Moth Larvae

Aquatic weeds use up nutrients in the water and so reduce the capacity of cultured areas to produce fish. Chemical, manual and mechanical control methods can be too expensive and biological weed control is ecologically acceptable and more dependable.

The Chinese grass carp (*Ctenopharyngodon idella* Val.) has proved efficient in controlling many common aquatic weeds. Some insects too may destroy particular aquatic weeds but their use for biological control has not received much attention. Philipose found that certain moth larvae were able to feed on the leaves of *Pistia stratiotes* L., a common noxious floating weed, but did not identify them or describe the infestation.

Here we report that the larvae of the moth, *Erastroides curvifascia* Hampson, feed specifically on foliar and stem portions of *P. stratiotes* (which is not eaten by grass carp or any other phytophagous fish) and describe their use as a potential biological control agent for the weed in fishery waters.

Field observations were made in a perennial pond (0.35 ha) with a dense infestation of *Pistia*. The numbers of larvae are high during the winter months (that is November to February) (water temperature: 22-26° C) and therefore the consumption of the host plant is also extensive during this period. The entire life cycle of the moth is completed within 27-35 d. The larval stage, which is the active feeding phase, accounts for more than 50 per cent of the life span. The late instar larva pupate in the deep furrows of leaves and petioles of *Pistia* after feeding for about 7-9 d. The pupae metamorphose into adults in about 5-6 d.

Three sets of preliminary trials were conducted in plastic pools (capacity 500 1; surface area 6,943 m^2). Field conditions were simulated by providing 50 kg of pond soil at the bottom and 2001 of pond water in each set. The weed density in the three plastic pools was identical, each receiving 1,650 g of fresh *P. stratiotes*. Two different densities of the late instar larvae, 86 m^2 and 17 m^2 of the surface area, were tried in the first and second pools respectively, the third being the control. The larvae in the first pool cleared 1,340g of weed in about 30d, and 1,230g of weed was cleared in 40 d in the second pool. The control set recorded an increment of 960 g in weed weight at the end of 40 d reflecting the growth of *Pistia* in the absence of predation by the moth larvae. It was observed that in both the experimental sets the larvae introduced initially pupated after a brief period of feeding, resulting in the emergence of moths which immediately bred. The new early instar larvae caused most of the damage to the plants.

Our study indicates the possibility of using the larvae of *E. curvifascia* for controlling *Pista,* as they largely satisfy the basic requirements of a potential control agent. The comparative ease with which the larvae can be transferred *in situ* to host plants and the way the moths readily breed in the laboratory suggest a wide scope for disseminating this important agent for the control of *Pistia* in fishery waters.

(Source: H. Chaudhuri and K. J. Ram; Nature 1975)

Chapter 13

Spread of Knowledge on Aquaculture: Prof. Chaudhuri's Effort Through Lectures, Discussions and Workshops

A RESUME

This chapter just outlines the subjects on which the professor or delivered lectures or discussed at various gatherings. It also gives a few front line aquaculturists names who have contributed to this science in India, south and southeast Asia and the world.

Indian and world aquaculture has reached a great height in technological discoveries due to success especially in the later half of 20th century. It should be very relevant to draw a broad outline of such success. In the present venture the editor has consulted the journals, *"Fishing Chimes"*, *"World Aquaculture"*, *"Aquaculture Asia"*, *"NACA"* and a few others in support of the editor's design. But most exciting has been the recovery of Prof. Hiralal Chaudhuri's handwritten day to day diaries from 1972 to 1996 (uniterrupted) which gave valuable information about his involvement in various facets of activities and development of aquaculture.

The editor draws the attention of the readers to the book entitled *"Aquaculture Beyond 2000: New Horizon"* edited by him on the technical work of the professor. Various chapters of the book deal with a wide range of subjects concerning aquaculture based on his work and experience in South and Southeast Asia which also record references of work done by other scientists.

The editor consulted the daily diaries of Prof. Chaudhuri and could gather information about his direct involvement in various facets of aquaculture from discussions, lectures and official engagements some of which are listed below:

A. Spread of Knowledge on Aquaculture through Lectures, Workshops, and Writings

- ☆ Aquaculture in Southeast Asia and far east
- ☆ Freshwater Fish Farming in China and India
- ☆ Aquaculture practices and extension in Thailand
- ☆ Freshwater Fish – A global food potential
- ☆ Prospects for world fisheries in 2000 and beyond
- ☆ Management of Aquaculture Fisheries
- ☆ Present status of Aquaculture in India
- ☆ Seaweed Culture
- ☆ Methods of control of aquatic weeds
- ☆ Aquatic angiosperms
- ☆ Fish farmers training in composite fish culture
- ☆ Farm Fish ponds and management
- ☆ Structure of the Pituitary and Thyroid of *Chanos chanos*
- ☆ Reproduction of Fishes
- ☆ Seasonal variations on the growth hormone
- ☆ Biochemical studies of the contents of pituitary of ovary of *Cirrhina mrigala*
- ☆ Photoperiodism in Fishes
- ☆ Abnormalities of the Gonads of Carp
- ☆ Techniques of hypophysation of Cat fishes
- ☆ Hormonal regulation of the Sexual cycle
- ☆ Applications of endocrinology to fish culture
- ☆ Growth, malnutritions and fecundity of cultivated fishes
- ☆ Induced breeding of *Mugil cephalus*

- ☆ Aquaculture and global seed production
- ☆ Hormones and induced breeding of cultivated fish
- ☆ Studies on the induced breeding of Milk Fish
- ☆ The theory and practices of induced breeding in Fish
- ☆ Early development, growth and maturity of *Cirrhina reba*
- ☆ Genetic selection and Hybridization of Fish cultivation
- ☆ Carp hatchery and nursery technology
- ☆ Aquaculture as an integral part of the Agricultural farming
- ☆ Fish culture in Paddy field
- ☆ Paddy-cum-brackish water aquaculture in coastal saline soil
- ☆ Fisheries potential of Air-breathing Fishes in India
- ☆ Penaeid shrimp culture in Asia
- ☆ Farming of Prawns and Shrimps
- ☆ Prawn and fish culture for increased yields
- ☆ Prawn feeds and feeding
- ☆ Prawn culture
- ☆ Seed productions and problems
- ☆ Running water culture system
- ☆ Design and construction of Fish pond
- ☆ Warm water aquaculture
- ☆ The Carps: Polyculture and Monoculture
- ☆ Selection of Fin fish for Aquaculture
- ☆ Food production through aquaculture
- ☆ Biological productivity
- ☆ The impact of aquaculture development on its environment
- ☆ Fish seed production and problems
- ☆ Marine fish culture
- ☆ Survey of edible oyster fisheries
- ☆ Pearl oyster fisheries
- ☆ Training on fresh water prawn farming
- ☆ Training on Crab breeding and culture
- ☆ Introduction of exotics

- ☆ Training on diagnostics of Shrimp diseases
- ☆ Trout farming in India
- ☆ Maintenance and management of Fish Aquarium
- ☆ Reservoir Fisheries
- ☆ Pond fish culture
- ☆ Use of probionics in Aquaculture
- ☆ Cage culture and its uses in India
- ☆ Aquaculture in maritime states
- ☆ Polyculture and integrated fish farming
- ☆ Parastic diseases of fish
- ☆ Training programme for farmers
- ☆ Tank management projects
- ☆ Running water culture
- ☆ Hatchery technology

B. Tilapia Farming

Culture of Fish Food Organism for Inland Aquaculture

- ☆ Biochemical studies in Rohu
- ☆ Pollution problems in Aquaculture
- ☆ Fish pathology
- ☆ Advanced Aquaculture
- ☆ Biodiversity and environment
- ☆ Intense carp culture
- ☆ Shrimp hatcheries and management
- ☆ Waste water aquaculture

C. Some Indian Fishery Scientists who Contributed to Aquaculture in India and South, Southeast Asia

The editor names a few (it is not possible in this treatise to name all) renowned scientists besides Prof. H. Chaudhuri, who have contributed fruitfully to build up aquaculture science in India and abroad. They are:

- ☆ Dr. S. L. Hora

☆ Dr. H. S. Rao
☆ Dr. A. V. Natarajan
☆ Dr. G. G. Jhingran
☆ Dr. M. Sinha
☆ Dr. A. G. Jhingran
☆ Dr. V. R. Pantulu
☆ Dr. H. P. C. Shetty
☆ Dr. T. V. R. Pillay
☆ Dr. V. V. Sugurnan
☆ Dr. T. J. Job
☆ Dr. B. S. Bhimacha
☆ Dr. S. P. Ayyar
☆ Dr. K. K. Vasu
☆ Dr. K. H. Alikunhi
☆ Dr. S. B. Singh
☆ Dr. R. D. Chakraborty
☆ Dr. S. D. Tripathy

Fishery Scientists who have also Contributed Substantially

☆ Dr. K. K. Sarojini
☆ Dr. Apurba Ghosh
☆ Dr. V. Gopala Krishnan
☆ Dr. M. Yusuf Kawal
☆ Dr. T. Rajyalakshmi
☆ Dr. S. D. Tripathi
☆ Dr. P. Das
☆ Dr. M. V. Gupta
☆ Dr. K. Raman
☆ Dr. C. S. Singh
☆ Dr. P. V. Dehadrai
☆ Dr. V. R. P. Sinha
☆ Dr. S. Ayappan

- ☆ Dr. N. C. Datta
- ☆ Dr. Raghu Prasad
- ☆ Dr. G. N. Saha
- ☆ Dr. P. R. Sen
- ☆ Dr. N. G. S. Rao
- ☆ Dr. S. R. Ghosh
- ☆ Dr. D. S. Murthy
- ☆ Dr. S. Jena
- ☆ Dr. R. D. Rao
- ☆ Dr. K. Jena
- ☆ Dr. D. K. Chatterjee

Chapter 14

Prof. Chaudhuri's Project Successes in a Few Countries

A RESUME

This chapter presents Prof. Chaudhuri's five project successes in LAO PDR, Nigeria, Myanmar (Burma), Fiji and Malaysia in a broad outlines.

Prof. Chaudhuri worked in LAO PDR for 5 years from 1979 to 1984 as Chief Technical Advisor and Coordinator on an FAO/UNDP project. An elaborate report was submitted to FAO/UNDP.

Prof. Chaudhuri made short and long term consultancies in several countries most of which have been referred to in the treatise entitled *'Aquaculture Beyond 2000: New Horizons'* published earlier.

In this chapter the editor presented a few short term projects successes in:

    (A) LAO PDR

    (B) Nigeria

    (C) Myanmar (Burma)

    (D) Fiji

    (E) Malaysia

Only the salient points of the projects have been brought to light.

Besides, he made two short term consultancies. The present one is on a short term project as follows:

(A) Applied Research Investigation and Technology Development in Fish Culture for LAO PDR (A short term project from July to Sept. 1986)

Rehabilitation of Fish Farm and Fish Culture Development

1. Terms of Reference

The terms of reference of the consultancy were as follows:

- ☆ To advise and coordinate in organizing basic facilities for applied research at the Nong Hai Centre (Vientiane).

- ☆ To participate in studies on breeding, growth, food and feeding habits of some important cultivable Mekong carps.

- ☆ To advise on future planning of the applied research programme in the country and in-depth studies on integrated fish/livestock/crop production systems.

2. Status of Aquaculture Research in LAO PDR

The level of aquaculture research in the country is at a very rudimentary level both with regard to available infrastructure facilities and qualified manpower. Under these limitations, some basic research physical facilities were established at the Nong Hai Centre to conduct simple replicated experimental trials on breeding, growth, food and feeding habit studies of Mekong fishes and evaluation of fish toxicants of plant origin for clearing ponds, etc. A set of 12 fibre glass pools (0.2 m, ht 0.75 m) with turn pipe arrangements were installed with facilities for water supply and aeration.

In addition to the above yard facilities, a laboratory room has been organized at the Livestock Directorate, equipped with aquaria, microscopes and with accompanying simple facilities for laboratory studies on the above aspects. Majority of the common glass wares, chemicals, general fittings have been already procured on the Consultant's advice, and items like electric balance, pH comparator, etc. are being procured.

The counterparts, already trained earlier in general fish culture activities, were constantly involved in learning the methodology and data collection for the above stated applied research investigations.

3. Applied Research Studies on Various Important Aspects of Fish Culture

The applied research studies were undertaken in the following order of importance after discussions with the Project Manager,

3.1 Evaluation of cultivable qualities of some Mekong fishes.

3.2 Preliminary growth studies of some Mekong fishes.

3.3 Induced breeding experiments: trial with new hormones.

3.4 Observations on the maturity and fertility of the hybrid Pa phone x Rohu.

3.5 Some observations on fish toxicants of plant origin.

In addition to the above, to familiarise the research counterparts with drawing up of a Research Proposal and to facilitate: proper recording and maintenance of research data, suitable formats with necessary explanations and instructions were prepared and appended to the report, which are listed below.

(*a*) Preparation of a Research Proposal.

(*b*) Study of Food and Feeding Habits of Fishes.

(*c*) Assessment of Fish-food Organisms (plankton) in Ponds.

(*d*) Recording Basic Data on Grow-out Ponds in a Fish Farm.

(*e*) Weekly Study of Essential Physico-Chemical Parameters in Fish Ponds.

(*f*) Growth Study of Fishes,

(*g*) How to Conduct Toxicity Experiment with Piscicides?

(*h*) Fecundity Study of Fishes.

(*i*) Maintenance of Records of Induced Breeding Experiments.

(*j*) Manuring Programme for Fish Ponds.

(*k*) Magnification Details of the microscope used for Research Studies.

3.1 Evaluation of the Cultivable Qualities of Some Mekong Fishes

Mekong river and its tributaries are quite rich in fish fauna and harbour a large number of commercial species, some of which could be suitable for fish culture in ponds along with the Asiatic carps and other cultivated species.In view of the above, preliminary studies were initiated with some of the Mekong fishes to determine their suitability to fit in the poly-culture system of carps. Attempts were made to collect as many Mekong fishes as possible both alive and in preserved conditions for studying respectively the rate of growth as compared to the established species, and for gut-content studies to determine the food and feeding habits and their compatibility with other cultivated species. The following species have been studied:

1. *Cirrhinus microlepis* (Pa phone)
2. *Osteoohilus melanopleura* (Pa nok khao)
3. *Puntius gonionotus* (Pa pak)
4. *Morulius ohrysophekadion* (Pa phia)
5. *Labeo dyocheilus* (Pa va)
6. *Catla carpio siamensis* (Pa ka ho)
7. *Probarbus jullieni* (Pa eun)
8. *Puntius orphoides* (pa pok)
9. *Puntioplites proctozysorn* (Pa sa kang)
10. *Hampala macrolepidota* (Pa sou)
11. *Notopterus chitala* (Pa thong)
12. *Ompok bimaculatus* (Pa souam)
13. *Kryptopterus* sp. (pa souam)
14. *Mystus nemurus* (pa kot)
15. *Pangasius* (Pa souay) sp.

In addition to the above species some minor species such as *Cirrhinus lineatus* (Pa soi) and *Cyclocheilichthys repasson* (Pa khao i thai) were also examined.

For collection of materials for the study, the Consultant visited the reservoirs Nam Houm, Nam Souang and Nam Ngum and made some collections. Counterpart officers were instructed to visit fish landing centres of Mekong and its tributaries to collect fish specimens for the study. Fish markets at Vientiane and suburbs were visited and guts were collected from fishes which were cut for sale, for gut-content studies. Some experiments have been initiated at the Nong-Hai fish form on the growth study of a few species of Mekong carps transported from Luang Prabang and stocked along with Asiatic carps catla, rohu, mrigal and also the common carp for comparison of their growth.

The publication "Fishes of the Lao Mekong basin" was very helpful in identification of the local species. However, there is hardly any information available regarding the cultural qualities, rate of growth, food and feeding habits as well as breeding habits of the local fishes. For lack of time no visits could be made to the provinces where several of the important species are available. Also, since the breeding season of most of the Mekong carps has just commenced, appropriate size of stocking materials were not available for initiating growth studies. However, from the studies made on the limited available materials and some studies made previously on several species, some observations have been made but these are tentative and further research need be done for confirmation: (speciewise – 15 species study is not presented to reduce the volume of the book).

3.2 Preliminary Growth Studies on Some Mekong Fishes

Based on the food and feeding habit studies and limited availability of fingerlings and juveniles of local fish species, a preliminary experimental trial on growth studies was initiated at the Nong Hai centre involving the counterparts. During the short consultancy period, planning of simple growth studies was explained to research technicians, and in fibre glass circular pools (0 2m, ht 0.75m), stocking of the local species *Puntius gonionotus* (pa pak), *Puntius orphoides* (Pa pok) and *Cirrhinus lineatus* (Pa soi) was done in duplicate set up. The above species were stocked in the pools along with known cultivable carps *viz.*, *Catla catla* (pa catla), *Labeo rohita* (Pa rohu), *Cirrhinus mrigala* (Pa mrigal) and *Cyprinus carpio* (Pa nai) at comparable number and size and also singly in separate pools Conventional supplementary feed (rice bran) and carbon manuring with liquid manure were given daily to the experimental pools in equal quantities).

3.3 Induced Breeding Experiments: Trial with New Hormones

The Consultant brought with him a small quantity of a new highly effective ovulating agent, the synthetic analogue of the nonapeptide LH-RH (Luteinizing Hormone Releasing Hormone) in combination with a chemical called Pimozide. This combination had been tried by some Canadian scientists with remarkable success and also successfully conducted on milkfish and seabass at the SEAFDEC Aquaculture Department at Tigbauan, Iloilo, Philippines. The LH-RH analogue can be injected intramuscularly or introduced into the peritoneal cavity as quick-shot pellets. Unfortunately, by the time the Consultant joined the consultancy at Vientiane, the brood fishes at Nong Hai Fish farm had mostly lost prime gonadal conditions and had passed the ideal stage for induced breeding experiments. This year, the rainy season had commenced quite early in Nong Hai area and there had been continuous heavy rain fall in May-June which caused early maturation in spawners and by July the brood fishes were loosing condition and the gonad had started resorption.

However, the breeding experiments were taken up shortly after the Consultant's arrival. The first experiment was initiated on July 10, 1986 when one set of catla was injected. The female spawner of catla was injected with mg/kg LH-RH in combination with 10 mg/kg of Pimozide in 6 per cent salt solution. Only one injection was given. The males were injected with common carp pituitary gland extract. Within 12 hrs. profuse breeding of Catla took place and over 50 per cent of the released eggs hatched out. This was the first successful spawning of this important Indian carp at Nong Hai farm.

On July 19, 1986 in another experiment, a gravid Pa phone-rohu hybrid female was successfully induced to breed by LH-RH and Pimozide combination. The only male available and introduced along with the female although was not readily oozing was given a low dose of the same hormone. The released eggs were all unfertilized and when the male hybrid was stripped, no milt came out. However, some freshly stripped eggs from the spawned female hybrid were fertilized by the milt obtained from one rohu male which were fertilized and a few hatched out.

3.4 Observations on Maturity and Fertility of the Hybrid Pa Phone x Rohu

Pa phone (*Cirrhinus microlepis*) is a popular Mekong carp and Rohu (*Labeo rohita*) is the tastiest Gangetic carp already introduced in Lao cultivable waters. While both the carps are rated high for their taste, but pa phone is a slow growing species and rohu a fast growing one. Thus the hybrid produced is expected to be an improved indigenous variety.

While detailed studies on the hybrid Pa phone x Rohu appear necessary when proper research facilties are developed to evaluate its cultivable qualities, the basic information on maturity and fertility of the hybrid are essential. The hybrid has been found to mature in ponds quite like other major Asiatic carps, the males being ripe in the second year and the females in the third year. The fertility of the hybrid has been established in the induced breeding trials during the consultancy period, as described under the Induced Breeding Experiments. Further, the fecundity of the hybrid was also assessed from a female hybrid (1 kg size) which was approximately 50,000 eggs/kg body weight, which is considered quite good.

However, a word of caution is recommended that the hybrid should not be propagated in cultivable as well as open waters until proper detailed growth, food and feeding habit studies are carried out. The apparent advantage of the hybrid is its expected faster growth rate than the preferred Mekong oarp (Pa phone) with its similar taste. Also, the hybrid being more in body depth compared to the father (Pa phone), it would have greater flesh for a. unit body length. The information that the hybrid is a fertile one would have considerable importance in evolving an improved variety through selective back-crossing.

3.5 Some Observations on Fish Toxicants of Plant Origin

It is important to locate and evaluate some indigenous plant fish toxicants in order to advise the farmers in preparing the fish ponds to achieve high survival rates of baby fishes. The bark of the plant toxicant Albizzia procera (ton thon), screened under the previous project LAO/78/014 was found to have saponin which affected fish kills at 100-300 ppra (crude bark basis), was re-evaluated. It has been found that the bark on storage loses its efficacy and the fresh bark alone appeared effective. Hence, its large-scale field application does not appear so positive. However, a sample of bark (2 weeks old) has been given to a pesticide laboratory in Bangkok to analyse its saponin content in order to collect additional information on this material.

With another possible angle of study to evaluate the comparative efficacy of the reportedly indigenous derris root material available in Luang Prabang, a commercial sample of roots was procured from Bangkok. The commercial root sample was thoroughly crushed and its toxicity to tilapia fingerlings was found quite satisfactory between 4-8 ppro, but the dose is likely to come down to 1-2 ppm when finely powdered derris root is used. The comparative study will be followed up when local derris material is obtained from Luang Prabang after the rains are over.

It is understood that seeds of *Strychnos nux-vomica* are quite commonly used by farmers in the province of Champasak as a fish toxicant and fishes so killed are consumed by the people without any obvious ill effects. Therefore, an adequate sample of the seeds of *Strychnos* was procured for toxicity studies. In the preliminary experimental trials, the scrapings of soaked seeds in water did not appear effective between 50-100 ppm. However, further trials may be given by properly grinding the seeds (seed coat is very hard) to determine the exact dose effective in killing fishes.

The available literature on mode of action of Strychnine, which is an alkaloid found in *Strychnos* seeds, that it acts primarily as a central nervous system stimulant. With this limited information gathered so far on *S. nux-vomica*, it is considered important to suggest that a toxicant having the above mode of action may not be very desirable as it may prove fatal or hazardous under careless handling. However, more trials with the toxicant material are envisaged by the project.

4. Future Planning of Applied Research Programme

The project LAO/82/014 has so far registered good progress in rehabilitating existing fish farms in the provinces of Charapasak, Savannakhet, Luang Prabang, Xieng Khouang and Sam Neua, and in addition developed another demonstration fish farm at Nong Hai in Vientiane Prefecture. Further, the low-input technology of integrated fish/crop/ livestock production systems has been developed at Nong Hai and Savannakhet centres for large–scale adoption throughout the country apart from facilities for general fish culture, fish breeding and pond management techniques at all the above centres. The Lao counterpart technicians have been trained abroad as well as on-the-job at respective centres and also through in-service training within the country. As a logical sequence of development in the field of aquaculture, simple applied research activities have to be initiated at national level in order to impart research aptitude to the trained counterpart technicians and critical thinking and evaluation and modification of the various aquaculture techniques to suit the needs of various parts of the country. This alone would lead to self-reliance in national planning and appropriate extension of the technologies so as to reach its benefits to the farmers located in the cooperatives and villages.

Hence, the essential follow up actions needed to achieve the aforesaid objectives are:

☆ To develop the basic infrastructure facilities of a modern fish farm (at a suitable, well-located easily accessible site) with drainable experimental ponds, reliable source of water supply, area for yard (semi field) experiments and a small field laboratory.

☆ To equip adequately the fish farm/yard/laboratory with suitable equipments and supplies.

☆ To post the qualified manpower to implement and monitor various applied research (simple technology testing) programme, according to national priority.

☆ To further train the available trained manpower in advanced aquaculture and research methods.

It is understood that a proposal on establishment of a National Applied Research Tranining Centre is already under the active is already under the active consideration of the national government, UNDP and FAO. In order to avoid duplication, the relevant proposals for future shall be covered under the heading 'Recommendations'.

4.1　In-Depth Studies on Integrated Farming Systems

The integrated system of fish culture along with crop/fodder cultivation and livestock raising viz., piggery, poultry and duckery have been subjected to test at the Nong Hai centre and also Savannakhet centre to limited extent. The suitability of this integrated farming system has been successfully demonstrated and this low input technology is highly flexible with regard to the type and extend of integration which shall be possible in different parts of the country with widely varying climatic and geographical conditions.

However, it shall be essential at a later stage of development that the parameters of integration of fish/crop/fodder/livestock are more precisely defined for various geographical regions of the country. This would mean a balanced allocation of space (land or water) and resources (feeds and manures) for the above systems in a manner that these activities continue to be complementary rather than conflicting with each other. For this the in-depth studies on the depth studies on the integrated farming systems shall have to be conducted at the proposal National Applied Research and Training Centre.

It is important to allot adequate land area for farming cereal and fruit crops, fodder and facilities for raising cattle, pigs, poultry and ducks along with water area for fish culture. As an example, for manuring hectare of water area for fish farming, 1000-2000 chicken/500-800 ducks/30-40 pigs could be considered appropirate and these tentative figures will have to be revised according to the needs of the area (cooperative/village), demand of produce and availability of input resources.

4.2 Recommendations

The applied research needs for aquaculture development in LAO PDR were carefully evaluated and to support the fish culture extension programme on a sound basis, the following recommendations are considered important by the Consultant.

1. To impart proper status to fisheries development in LAO PDR, a separate Department of Fisheries may be established under the Directorate of Livestock and Veterinary Services. The Department of Fisheries should undertake full responsibility for planning and implementation of fisheries research and development programmes for the country in accordance with the national priorities. Under the above stated department a National Applied Research and Training Centre may be established for conducting investigation and training on integrated fish/crop/livestock production systems. It is advisable to select a well-located site for the centre. An Experimental Fish Farm and a Field Laboratory should be established with modern, but simple facilities for applied

fish culture research studies. This establishment would serve as the nerve centre for simple applied research, training and extension programmes both at the Central and Provincial Governmental levels.

2. As far as possible, the existing infrastructure facilities already established by the various developmental projects on fish/crop/livestock production systems should be utilized in order to avoid duplication of expenditures and resources.

3. The other major need is to impart long-term (2-3 year) advanced training in aquaculture research methods to the qualified fishery technical officers in order to achieve self-reliance in development of national manpower.

4. The National Applied Research and Training Centre should have a Documentation Section to compile all available data and information on fisheries in particular and other related disciplines in general.

5. Following areas of applied research appear significant as follow up programmes:

   (*a*) Continuation of studies on growth, food and feeding habits of the most promising cultivable Mekong fishes such as Pa nok khao, Pa phone and Pa ka ho. Out of the above speices, Pa nok khao may prove to be and excellent addition to the multi-species culture system in view of its voracious phytophagous habit like the grass carp.

   (*b*) Breeding habits of the selected Mekong fishes viz., Pa nok khao, Pa phone and Pa ka ho may be studied and some trials on their seed production may be conducted, if possible, under the existing facilities,

   (*c*) Some detailed studies on the Pa phone x Rohu hybrid should be carried out when an experimental fish farm is established.

   (*d*) Greater search for locally available fish toxicants of plant origin should be continued and their efficacy evaluated. This would serve as an important support to the fish pond management aspects.

4.3 Additional Support to the Project

Apart from terms of reference provided for the consultancy, additional support was provided to the Project Manager in strengthening the ongoing training and extension activities. These are:

1. Finalization of the Fish Farmers Calender

2. Preparation of transparencies on various aspects of fish culture as aid to the on-going in service training programme at Nong Hai, Vientiane.

3. Delivered a talk to the counterpart technicians on general fish culture practices.

(B) Aquaculture Consultant in Nigeria (Africa)/Ellah Lakes Limited Fish Farm Project in Nigeria

On request from International Aquaculture Consultants of Makati, Metro, Manila, Philippines, Prof. Chaudhuri acted as a consultant on a 10 days consultancy for a reconnaissance-cum-technical assistance at Ellah Lakes Limited Fish Farm Project, Obrikorn Rivers State Nigeria.

The report and recommendations testify to the high quality advisory role played by Prof. Chaudhuri when he was servicing as Senior Technical Consultant at SEAFDEC in 1988. Some excerpts of his recommendations in his own words are presented which are sure to testify to his indepth grasp of various facets of aquaculture in general; and also how people harvested the fruits from his experience especially on commercial production of fish not only in Asia but also in other continents.

1. Training

The consultant attended a seminar where he talked on the basic principles underlying fish production, nursery and hatchery management, pond fertilization, mono and polyculture of fishes, and on integrated crop-fish-livestock farming systems and recycling of wastes. Twenty three technical staff attended and most of them showed their genuine interest and asked relevant questions. In the absence of an overhead projector, colored slides were shown to them. Ten published papers of the Consultant on induced breeding and nursery management were furnished for the use of the farm library and his book on induced breeding of carps and another reprints were kept by the Project Manager for xeroxing and return.

In general it is felt that the staff have adequate background for the work. However, on-the-job training on various aspects of pond management will be needed.

2. Prospects of Continuing Assistance from IAC

The consultant discussed with the Chairman regarding the assistance (long-term or short-term) which he expects from IAC and was told that "Although present management/production staff have the requisite know-how and special skills required by the project, most of them have not had sufficient practical experience in commercial production of fish." That is why he has been requesting for technical advice on collaborative basis with foreign companies with a track record in the field. He considers that such special arrangements with foreign companies who have had decades of experience in this field would help make double assurance of success in the project. The advice and cooperation envisioned will be mainly on production methods. Hence when talks about the remuneration of the proposed consultant came up he said and also wrote IAC that "The expatriate we want would be around here in the capacity of a special advisor who can ensure that the hatchery and the grow-out pond operations are properly carried out in order to achieve the best result". He also stressed, "For instance, we would require a guaranteed production of 10 to 20 tons per ha in order to justify our investment".

3. Summary and Recommendations

Consolidation of Existing Infrastructures

A preliminary assessment of the present status and functioning of the Obrikom fish farm of Ellah Lakes Ltd was made and it was observed that a lot of work has yet to be done to complete the facilities to bring the entire fish farm under full production.

It is recommended that whatever infrastructures have so far been established (Phase I and II), should be consolidated first before venturing into Phase III and Phase IV of the project. This is absolutely essential in view of the huge investment that has already been made. Many pond embankments are being eroded, several got submerged by flood water and about a dozen of the ponds have unfinished embankments. All these need immediate attention. The ponds adjacent to the store and office are reported to be giving better results than the distantly located ones. This indicates that the latter ponds did not get proper attention probably because of the distance and difficulties of reaching them. Hence for proper supervision and quick transport of feeds, fertilizers, seeds, etc. the approach road to each and every pond should be repaired and compacted to prevent erosion so as to facilitate easy and quick movement of vehicles. So far only about 50 per cent of the fish farm area could be stocked and others are still lying fallow.

4. Aquaculture Engineering

Although the company has a foreign trained qualified Civil Engineer as the Project Engineer, it would be worthwhile to reinspect the entire layout of the fish farm (when the flood water recedes) with the help of a qualified and experienced Aquaculture Engineer who may be able to suggest improvements in the layout of the fish ponds, proper installation of the inlets and outlets, ways and means of maintaining free egress of any excess water by preserving as much as possible the original course of Ebiam stream flowing through some of the fish ponds and also providing a suitable layout of the new ponds to be constructed in the near future–Also, said engineer may be able to advise construction of a reservoir and channeling water from the stream and also from the springs. Since there is a gradation of over 7.5 m, the water from the reservoir could possibly be fed to all ponds situated at lower levels by gravity. The reservoir may include the 15 ha pond and 8 ha swampy area to hold sufficient water for feeding even perhaps Phase III ponds when constructed. At present, Pond No. 44 which has been designated as a reservoir is less than 1 ha in area which is apparently too limited. Besides, to prevent flooding of the ponds adjacent to the swamp and Ebiam stream, an escape channel carrying additional water of the stream should be constructed if the engineer thinks it would be useful and feasible.

It appears that the planning of the fish farm based on the topographical survey of the area and construction of the ponds have not met the requirements of an ideal productive fish farm. The blue-print of the ponds of Phase I and II provided, does not represent the actual pond size and shape and the location is quite different from the present set *of* ponds in several places. Besides, the soil texture and permeability of soil probably was not thoroughly studied taking proper samples from the different areas of the fish farm. A good number of ponds are constructed on porous soil which does not hold any water.

5. Topographical Survey and Study of Soil

It is recommended that before taking up further construction of ponds for the Phase III and IV of the project, detailed topographical survey of the proposed pond area be made and both physical arid chemical characters of the soil studied. Soil having clay, silty-clay-loam, sandy-clay texture are usually impervous and hold water. Vertical profile of the soil up to 1 m depth or even deeper in selected places and samples collected from the different areas of the proposed fish farm should be studied as to its permeability, and analysis of soil for pH, available N, available P, total N and organic matter content determined. It would be worthwhile to study the soil samples of ponds of Phases I and II which probably have not been done very thoroughly before accepting that the soil as reported is silty-clay (25-65 per cent clay).

6. Hatchery-nursery Complex

The present hatchery-nursery complex is estimated to produce 5 million fingerlings per annum. To produce 5 million fingerlings and at the present low rate of survival of hatchlings and fry, 25 million post-larvae have to be produced. To rear these, the present nursery and outdoor tank: space is not adequate. Also to obtain these post-larvae, a broodstock of at least 2,000 pairs of male and female brooders (500 g wt) would have to be developed.

While induced breeding of *Clarias* by hormone injection as practiced by the hatchery technicians is quite successful, the same is not true with the common carp breeding. A production of a meagre 24,000 fingerlings of common carp have been achieved in one year. The broodstock development, breeding and hatching techniques and rearing of fry of common carp need much improvement in Clarias too, hatching of eggs and larval rearing methods have to be improved for achieving higher survival of seed at every stage.

The kakabans or egg collectors used for Clarias and common carp breeding are not adequate. It is suggested that the present material may be substituted by finer materials if available. The kakabans, preferably in square bamboo frames, should be distributed in the breeding tank to cover at least half of the water surface. This will largely prevent shedding of loose eggs which drop down to the tank bottom. Besides, the present breeding tanks are too deep which hinders proper observation and handling of breeders and collection of kakabans. The outside tanks, however, have proper depths.

The mortality of Clarias and carp fry can be considerably reduced by proper feeding, aeration and with better technique of handling. While rearing Clarias fry, periodic segregation of different size groups of fry and early fingerlings should be done by sieving through screens of different mesh size so as to reduce cannibalism among the fry.

It is recommended that zooplankton such as rotifers and cladocerans may be produced in mass scale by manuring with fresh cowdung/poultry droppings/ pig manures. These zooplankters should be fed to the postlarvae of Clarias and carp from the third day after hatching. This may be followed by simultaneous feeding with finely powdered supplemental feed. Once the technique of mass culture of cladocerans (*e.g.*, *Moina*, *Daphnia*, etc.) is developed, the same could replace Artemia which is very expensive and involves foreign exchange. Similarly, regular collection of pituitary glands from mature common carp grown in the farm will save considerable expenditure and foreign exchange. Detailed instructions on all the above techniques were given by the Consultant at site and relevant literature handed over.

Besides, natural fishfood organisms such as earthworms, tubificids and bloodworms (Chironomid midge larvae) could be cultivated for feeding Clarias fry and fingerlings which will appreciably reduce cannibalism in the early fingerling stage and increase their survival and growth rate.

It is recommended that the quality of water supplied to the hatchery should be studied as to its pH; dissolve oxygen, free CO_2, hardness and total alkalinity. The water should be free from industrial, agricultural and other contaminations. The volume of water flow (at present only 3 1/min) to the hatcheries, should be increased. Further, the water should be free from Cyclops: a predatory copepod (micro-Crustacea) and insects (notonectids) and other harmful organisms. The spring water in Pond 63, if found suitable could be used. At present, water is occasionally pumped from Pond 1 (lake) which being a production pond is manured regularly.

Besides, water from the hatchery returns back to the supply pond and recirculated. This should be avoided since in case of any disease/parasitic infection in one tank the same may contaminate the entire hatchery.

The nursery management technique may be further improved. Complete control of predators including insects is essential. The larger insects, tadpoles, etc. can be removed by repeated netting with close-meshed nets and back-swimmers killed by application of oil-emulsion. The nursery margin (at water level) should be kept clean of grasses and weeds which harbour predatory insects and their larvae. Since the pH of water is around 6.5, the ponds should be limed adequately and heavily manured for production of sufficient zooplankton. Once fry is stocked, regular carbon manuring in the forenoon (as advised) be done to maintain the swarm of zooplankters.

Feedmill

Delay in the completion of the feedmill appears to be a bottleneck in the large-scale production programme of the project. It is recommended that efforts should be made to establish the feedmill at a very early date and start production of the pelletized fishfeed. In the absence of a formulated feed, at present spent grain alone is being fed. The feedmill is estimated to produce 3 tons of pelletized feed per hour. The feed formulations as proposed, comprise of components more or less similar to the dry pellets compounded at the Fish Culture Station, La Landjia, Bangui, Central African Republic for Clarias. This has about 40 per cent protein. Although the proposed feed contains 10-20 per cent oxblood (dry), the feed would lack some important ingredients since the major dietary protein components come from vegetable sources. Animal protein like fishmeal (which is a costlier item no doubt) contains some growth factor and has delicate balance of amino-acids which is lacking in soyabean or other vegetable cakes/meals.

It is recommended that the proposed feed formulations should be given a trial before going into large-scale production of the pelletized feed. Also, it is advisable that a qualified fish nutritionist study the feed formulations and gives his/her advice whether the feed needs any other essential ingredients, amino-acids or any additive like terramycin be added to it. Trials also should be given so as to determine the performance of the feed in water, fish growth and its staying quality. Quality control of the finished product will ensure that the ingredients were mixed in right proportion. However, an analysis of the various components may be done initially to determine per cent of crude protein, crude fat, minerals and gross energy contents of the feed. Very often processing affects the nutritive value of feed. Also, proper storage of the feed stuffs is very important. It is to be experimented whether the feed is acceptible to *Clarias*, carp and tilapia and good growth is obtained. Very often calories from oil-seed meals which come partially from carbohydrates are less digestible to warmwater fishes. All these considerations are essential for a complete feed. However, for supplementary feed, the feed formulation is much easier. Further, the feeding technique, intensity of feeding (per cent of bodyweight) and the time and interval of feeding has to be determined and the economics, feed conversion ratio, etc. studied.

7. Species Cultured

At present polyculture of three main species such as *Clarias lazera* (=ga-riepinas) *Cyprinus carpio* and *Oreochromis niloticus* (Nile tilapia), is being practiced with the addiction of a few specimens each of *Heterotis niloticus*, *Heterobranchus longifilis* and *Gymnarchus niloticus*. It is presumed that Clarias would control tilapia population effectively. But so far there has not been any study made to establish that *Clarias* can effectively control tilapia population in a pond. *Clarias* is not a piscivore; it is rather an omnivore mostly feeding at the bottom on insects and insect larvae, worms and other bottom organisms. It may feed on small fish occasionally or especially in the early fingerling stage. This aspect needs further investigation. In fact, one of the ponds stocked with *Clarias* and tilapia was found to have thousands of tilapia of different sizes. The grow-out pond corners should be occasionally netted with fine-meshed net piece (for tilapia fry) and 4.6 mm mesh net (for tilapia fingerlings) to check abundance of tilapia population and take remedial measures. Further, the ratio of each component of polyculture species should be experimentally determined.

Heterotis is a herbivore (as reported), quick growing species and may be included as one of the components of the polyculture species. It breeds in ponds and shoals of fry could be collected from the nests. Artificial breeding or development of broodstock in special ponds and providing nesting materials and facilities would helpful in the production of adequate fry for stocking.

Heterobranchus may compete for food with *Clarias*, but since the species grows much bigger (1 m and 20 kg), it may be cultured along with *Clarias*.

Gymnarchus (electric eel) is a piscivore, but has a smaller mouth and could be a good predator for controlling tilapia population. Its flesh is very popular and also breeds in ponds. However, mass production of its seed may be difficult at present and needs further experimentation.

Channa obscura, the snake-headed predator is not-a very popular fish. This should not be introduced into the farm since it would migrate to other ponds and take a toll of Clarias and carp fry and fingerlings.

Lates niloticus (Nile perch) can be tried as a piscivore for controlling tilapia population in ponds.

8. Production Management

In order to properly coordinate and consolidate production management of the fish farm operations, the Project Manager (PM) with the assistance of his technical staff should prepare and submit a detailed Operational Management Plan (OMP) for each forthcoming year. This should be done during the period between August to October– of the current year. The PM submits the OMP to the farm Board of Directors (BOD) for deliberation. The BOD approves as submitted, or with modification (s) as may be found necessary, said OMP including the necessary budget to carry it out on or about November of the current year. The PM and his staff make preparations for the implementation of the incoming year's OMP in December ready for full scale implementation come January.

Once the OMP is approved by the BOD the PM and his staff should be given full authority to implement said OMP. If any change(s) in this approved OMP is found necessary such modification(s) should first be approved by the BOD.

9. Additional Management Initiatives

To further enhance productivity of the ponds, the following possible additional initiatives may be undertaken.

(*a*) Introduction of Exotic Species

After a careful study and if it does not conflict with existing government policy, the company should consider introducing the quick growing Asiatic carps such as grass carp (*Ctenopharyngodon idella*) and bighead (*Aristichthys nobilis*). These are very valuable cultivable species and are capable of giving very high production. Besides, grass carp can be used as a biological control of noxious aquatic weeds as it feeds voraciously on grasses and other land weeds. Bighead can grow 4-5 kg within 4-5 months. In addition, silver carp (*Hypophthalmichthys molitrix*) and rohu (*Labeo rohita*) can be tried on later dates in polyculture. These actions on introduction of exotic species should also take into consideration the acceptability and marketability of the various species and the possible introduction of diseases.

(*b*) Integrated Crop-fish-livestock Farming System

It is suggested after some study on acceptability of the product that the company may start integrated farming initially at a small-scale which could be expanded later on if found profitable. Although organic manures (poultry droppings) are reported as available, to be self-sufficient and also to save expenditure in transportation, poultry keeping could be integrated with polyculture especially when quick growing Asiatic carps are introduced. The spacious land area could be utilized for crop farming. Integrated farming would give high production and would help in making the farm self-sufficient in its needs.

(c) Running Water Fish Culture

The presence of the Ebiam stream and springs in the farm gives some opportunity of utilizing the flowing water for running water fish culture. On an experimental basis running water culture with catfish and carps could be started in the ponds which do not hold water and have heavy seepage. With intensive stocking and intensive feeding with complete pelletized feeds, production from running water fish culture is expected to be very high.

10. Training of Staff

The present Acting Project Manager and other Technical Staff are well qualified and appear to be carrying out the project activities quite satisfactorily but majority of them lack sufficient practical experience in the specialized techniques of broodstock development, induced breeding, hatchery and nursery management and in various other aspects of pond management. It is recommended that a selected few of them be given practical training in aquaculture management. Although the technical staff have been given some sort of general training to take some preventive or curative measures in case of some common parasitic infections and common diseases, it is advisable that at least one technical staff be given adequate training in fish pathology and should be able to tackle any immediate problem of fish mortality in the hatchery as well as in grow-out ponds. Also one or two technical staff should be trained to monitor water quality such as pH, dissolve oxygen, CO_2, total alkalinity, etc.

11. Procurement of Essential Field and Laboratory Equipment

While the hatchery laboratory is equipped with microscopes, glasswares, chemicals and other equipment, some essential cheap and useful small items such as small plankton hand nets, droppers/pipettes, covers lips, grooved slides, small white enamel trays, small (assorted) sieves (like tea-sieves with handle), needles, etc. are lacking. Also, a 50 mm objective for the microscope would be very useful in examining developing eggs as well as bigger zooplankters.

It is recommended that paddle wheels (electrically operated) or other aerating facility be installed in selected ponds for aeration purposes. With the target of achieving 20 tons/ha/annum using heavy stocking and intensive feeding there is the risk.of oxygen depletion and large-scale mortality of the stock. Aeration by a paddle wheel or other means will eliminate this risk of fish kills by O_2 depletion and also would help increase production by eliminating some of the metabolites which may inhibit fish growth.

12. Marketing of Products

As regards marketing of the products, at present there is hardly any problem since the demand of fish is very high and fishes are readily sold at the pond site or at the local market. The area is very densely populated and there is heavy demand for fish in the neighboring towns and cities. However, in future when the company starts full production aiming at as much as 20 tons/ ha/year the marketing may not be as easy as it is now. It is recommended that the company may in near future make a thorough study of the capacity of the market and the methods of transportation, post-harvest handling and distribution of the products.

With the completion of the Phase III and Phase IV of the project, the annual production target is estimated by management to reach around 8000 metric tons of fish and 400 metric tons of prawns. The company should then be equipped with an ice plant/cold storage with more elaborate arrangement for marketing. With the larger production it may be necessary to take up processing of finfish too in addition to the proposed processing plant for prawns.

13. Economic Analysis

Ellah Lakes (Fish Farm) Ltd has prepared a good report supported by several tables on the cost analysis for the 100 ha fish farm incorporating project recurrent costs (for 10 years), cash flow and profit revenue. It would be advisable for the company to make an Economic Feasibility Analysis of the project which would receive considerable investment. Although such aquaculture projects are run primarily along profit-oriented lines, they also contribute to the socio-economic development not only to their community but also of the entire economy of the area contributing to improvement of the standard of living, community development, foreign exchange earnings, lowering of prices, and the utilization: of locally available materials.

14. Protection of the Fish Farm from Floods

It is strongly recommended that some effective measures should be taken to protect the entire fish farm and other facilities from floods. Although construction of a flood proof perimeter dike would be very costly, effort should be made to protect the ponds from regular floods either in totality or by portions.

(C) Burma Project

Since the success in induced breeding of Carps in 1957 FAO/UNDP/TA a useful 83 days consultancy in 1966 for development of Fish Culture in Burma.

Pror. Chaudhuri's work impressed the then government of Burma so highly that they sought his extension. FAO/UNDP/TA on Govt. of India's approval extended the period of consultancy from 1967 to 1970. Elaborated report on the consultancy was submitted to FAO/UNDP/TA.

This presentation is a summary of Prof. Chaudhuri's 83 days short term work in Burma in 1966. In fact, the Govt. of India received a large number of request for release of the professor for consultancy since his success in induced breeding but he did not get release from his government.

This project originated when the Govt. of Burma made a request to FAO for technical assistance in induced breeding of fish. This was to advise and assist the Govt. in

- ☆ Development of fish culture with special attention to the breeding of cultivated cars in ponds
- ☆ The selection and rearing of brood fishes
- ☆ Techniques of pituitary gland injection
- ☆ Hatching and nursery management
- ☆ Training the local workers

Some excerpts from his work are reproduced which would be of immense interest to the breeding aquaculturists and register how South and Southeast Asia was benefitted from his work.

1. Training in Techniques of Induced Breeding

Almost immediately after arrival, the expert started his work at the Hlawga Fisheries Station situated 14 miles from the city Rangoon.

Fish were examined from ponds previous stocked with major carp breeders. The majority of the breeders had completely resorbed their gonads, and even the roughness in the pectoral fins in most of the males had disappeared. Luckily, however, one female rohu (*Labeo rohita*) in fairly good condition was obtained. As no freely oozing (running) male was available, one male, though not in an ideal condition, was selected for the experiment. The pair were injected with fish (homoplastic) pituitary hormones. The female ovulated successfully releasing about one million eggs. Only a small percentage, however, were fertilized and these died in the course of development, probably because of over-congestion in a small hapa with innumerable fertilized eggs resulting in depletion of oxygen. Further attemts to collect suitable breeders here were unsuccessful.

Twante Farm was next searched for breeders but at this farm also, gonads had been resorbed in most of the breeders examined. Only one female *Labeo gonius* and two female *Labeo pangusia* with egg-distened abdomens were collected. No oozing males could be caught. The breeders were transported to Hlawga Pish Farm where a male specimen of *Labeo rohita* in fairly good condition was available. The female *Labeo gonius* died after injection, and her ovarian eggs when examined were found to be in the process of resorption. One of the two *Labeo pangusia* also had gonads in the resorting stage. The other one was in fairly good condition.

In the absence of a male of the same species hybridization was tried. The female *Labeo pangusia* was injected with pituitary hormones and when the fish was ready for stripping, the eggs were stripped into a plate and fertilized by sperm from an injected male *Labeo rohita*. A second lot of eggs was also artificially fertilized by milt from a male *Cyprinus carpio*. The development of the hybrid eggs proceeded normally till the embryonic streak was formed, but curiously enough, development then stopped in all eggs whether kept in trays in the laboratory or in hatching hapa in a pond. The experiments had to be stopped because no more breeders were available.

However, despite these difficulties, it was possible to give thorough training to the counterpart officer and other co-workers on the various operations in the technique of induced breeding of carps by pituitary hormone injections, including the collection and preservation of pituitary glands, method of recording weight of glands, preparation of gland extracts, rearing and selection of proper breeders, identification of sex, calculation of correct dosages, method of injection and hatching techniques. Training was also given in and demonstrations made of methods of artificial fertilization of carp eggs and hybridization.

2.   Survey of Fish Seed Production

With a view to determining the quality of fish "seed" from different collection centers in the River Irrawaddy and its tributaries, the expert visited centers in both lower and upper Burma.

Spawn Collection

The Directorate of Fisheries, Burma, has made extensive surveys in the River Irrawaddy and its tributaries and in the River Sittang and has established the following spawn collection centers.

Sl.No.	Center	River
1.	Kyawkmyang	Irrawaddy
2.	Sagaing	
3.	Mandalay.	"
4.	Ava	Miyintge
5.	Shwedaung	Irrawaddy
6.	Tharrawaw	"
7.	Henzada	"
8.	Tandoon	Pan-Hlaing
9.	Pantanaw	Irrawaddy
10.	Shwegyin	Sittang

Since the spawn collection from these centers had been completed long before the arrival of the expert in Burma, he could not view the actual operations. Reports from field officers of the Directorate of Fisheries indicated that the quality of spawn in most of these centers is not very good. It was reported that a better percentage of economically valuable species is obtained at Tandoon and Tharrawaw centers than at the Mandalay and Sagging centres situated in the upper reaches of the River Irrawady. The collection from the Myintge River has been consistently very poor. Shwegyin, a recently established collection centre in the River Sittang gave quite a good collection of spawn but the quality was very poor. Examination by the expert of some collections of fry reared in the nurseries at Hlawga Fish Farm confirmed the above findings.

3.   Collection of Fry and Fingerling

Fry and fingerling collection centres at Kyawkmyang, Mandalay, Ava, Shwedaung and Yandoon were visited by the expert to study the species composition of the advanced fry and fingerlings of carps collected there. The percentage of major carps in general was not very high, and it was felt that the centres located in the uppor reaches of the River Irrawaddy had a greater percentage of minor carp species, namely *Labeo bata, L. boga* and *L. pangusia* formed the bulk of the catches in the former centres. At the Myintge River centre the catch consisted mainly of *Labeo bata* and a few *L. calbasu*.

From these observations and also from the experience of the field staff of the Directorate of Fisheries, Burma, it can be concluded that the spawn available in the rivers of Burma so far surveyed do not contain a very high percentage of economically use varieties of carps. The same was also confirmed by observations made on the juveniles of carps brought to the markets for sale.

4. Identification of Carp Fry and Fingerlings

An intimate knowledge of the field identification characters of fry and fingerlings of major and minor carps is essential for a fish culture worker in Burma. He should be about to identify them correctly for proper stocking as well as for distributing quality of fish seed to the public. The expert gave practical training in methods of identification of fry and fingerlings of various major and minor carp species in the riverine collection.

5. Lectures on Fish Culture

As requested by the Government, the expert delivered a series of lectures of fish culture, including the technique of induced fish breeding, for the benefit of the staff of Directorate of Fisheries and People's Pearl and Fisheries Board; post-graduate zoology students, B.Sc Zoology (Honours) and Zoology major (Ichthyology) students of the Rangoon University. some zoology lecturers and demonstrators were also in attendance. The following topics are included: reproduction of fishes; factors influencing spawning of carps; induced breeding of carps by injection of fish pituitary hormones; development of carp eggs; identification characters of eggs, fry and fingerlings; principles of fish production; nursery, rearing and stocking pond management; mixed fish-farming; and pond fertilization; artificial feeding; fish diseases and parasites; exotic fishes, and related topics.

6. General Appraisal of Fish Culture

6.1. Past Development

Fish culture in Burma is a recent enterprise and the interest for the development of this industry practically started after independence was attained.

Ling (1955, p.3) noted that:

"Neither stocking operations nor cultivation are practised, although for purposes of accuracy mention must be made of the holding of fish in some temple ponds for religious purposes. It is difficult to explain the absence of these practices in view of:

1. The presence of so many bodies of water of all sizes suitable for stocking and cultivation;

2. The availability of suitable species ["especially the Indian major carps] for these purposes; and

3. The cultural and technical contact between Burma and Bengal in the west and China in the north-east.

Since it is certain that the absence of these practices is not caused by any biological, physico-chemical or other natural circumstances, the conclusion is inescapable that certain social and economic factors have been responsible.'

One reason for this may be the religious belief of Buddhists which forbids the killing of animals and in purity of livelihood, did not encourage people to engage themselves in the industry. According to FAO/UN (1957, p.33), "The fishermen of Burma belong to the group with the lowest social status in the community....". This factor may also have been responsible for the people's apparent indifference towards this industry. Farther, it is probable that, until comparatively recently, the native fisheries based on the natural wild stocks, particularly those of the delta and seasonally flooded areas, together with the practice of preparing fermented sauces and pastes from local and seasonal surplus supplies of fish, more or less adequately served the needs of the population.

It was only after the country attained independence that a few interested persons especially in lower Burma started stocking some ponds and other small water areas with fingerlings of major carps collected from the River Irrawaddy or the adjoining fisheries. But due to lack of proper technical knowledge these attempts at fish culture did not meet with much success. Interest in fish culture received a fresh impetus however when the fish farm at Hlawga Fisheries Station was established by the Fisheries Division of the Agricultural and Rural Development Corporation (ARDC) and private fish culturists were supplied with fry of cultivated species and given the technical know-how of fish culture.

At present the interest is widespread throughout the country and many private persons are taking up fish culture as an industry more so because of the rising price of fish and the high profit made by fish farmers. Since the law forbids construction of fish ponds in lands suitable for paddy culture, large-scale construction of new ponds for fish culture has not materialized. However, fish culture activity throughout the country is increasing day by day, especially in Mandalay and neighbouring areas in upper Burma where fish farmers are constructing small ponds in any available space and culturing Tilapia. Consequent to the increase in popularity of fish culture, the demand for the supply of fish seed of the food fishes has increased tremendously and the Directorate of Fisheries is trying to meet this demand.

6.2. Government Fish Culture Stations

With a view to promoting development of fish culture and initiating research on its problems, the Government of Burma has established several fish farms and fish culture stations. At present there are six stations under the Directorate of Fisheries: two purely fresh water, two marine, one brackish water, and one a research station meant for research on both inland and marine fishery problems.

Hlawga Fisheries Station

This station, established in 1952, is 14 miles from Rangoon, below the Hlawga Lake dam which supplies water to Rangoon. The fish faro has a total area of 55 acres which includes 13 nurseries, 5 rearing ponds and 14 stocking ponds. There are also a few cement cisterns for rearing spawn. The farm gets its water from Hlawga Lake. The ponds have individual inlets and outlets and can be drained or filled independently.

Immediately after its construction the fish farm was used for Tilapia culture and as a Tilapia distribution center. From 1957 major carp culture was initiated at the farm and this is now the main center where the bulk of the carp spawn collected from rivers is reared and advanced fry and fingerlings distributed to public and governmental agencies. Supply of quality fish seed is one of the main activities of the fish culture section of the Directorate. Major carp breeders are being reared in a few stocking ponds to be used in induced fish breeding experiments during 1967 when this farm will be the main center for the commercial production of fish seed. Besides major carps, exotic fishes such as common carp (*Cyprinus carpio*), tilapia (*Tilapia mossambica*), gourami (*Osphronemus garamy*) and a few specimens of grass carp (*Mylopharyngodon pioeus*) are also present in the ponds.

Eight of the large-size stocking ponds having a total area of about 35 acres have been transferred to PPPB and are at present under the management of the Board who are mainly concerned with the production aspect of fish culture.

Phalan Research Station

The Phalan Research Station is situated in the Kyaukton township of Hanthawaddy District, about 13 miles from Rangoon. It was established in 1955, intended for carrying out research on inland as well as marine fisheries. Inland fishery research envisages problems such as those concerned with experimental culture of major carps and tilapia, pond fertilization, and culture of milkfish (*Chanos chanos*) and mullet (*Mugil* sp.). The station has eight ponds with an area of 3 acres for brackish water fish culture and 15 acres, comprising three nurseries and six stocking ponds, for fresh water fish culture. A 70-acre reservoir supplies water for the fresh water ponds.

Thayetken Inland Fisheries Station

Thayetkon Fisheries Station was constructed in the vicinity of Mandalay town in upper Burma in 1957. The fish farm consists of nine nurseries, three rearing ponds and one stocking pond, having a total area of 6.5 acres. Tilapia, as well as major carps, are cultured. The nurseries are used for rearing carp spawn for distribution of carp fry in upper and central Burma. One pond has been stocked with the exotic fish Sepat siam (*Trichogaster pectoralis*).

In addition to the above-mentioned fresh water stations, the Pyinpyaman Fisheries Station at Kyaukpyu is mainly used for the culture of milkfish.

Twante Fish Farm

Twante Fish Farm, under the management of the People's Pearl and Fisheries Board of the Government of Burma, is located 18 miles from Rangoon, adjacent to the Twante canal which is connected with the River Irrawaddy. The total area under cultivation is now 104 acres. This will be extended to 320 acres during 1966-67, and according to the plans of PPFB, a total of 1,000 acres of water area will come under cultivation by 1970.

The ponds, provided with inlets and outlets, receive water from the Twante canal which is connected with the River Irrawaddy and has tidal effects. The water is slightly saline.

The ponds are stocked with major carps and also with common carp (Israeli strain). The fingerlings of major carps are supplied by the Directorate of Fisheries. Usually two to three-inch fingerlings are stocked at the rate of 3,000 per acre. Artificial feeding is done daily with rioe bran. Poultry farming has been started on the farm to secure poultry manure for fertilizing the fish ponds. Super-phosphate is also used, though not very regularly. The ponds are usually fished out every two years. Fish production is reported to be roughly 1,800 Ibs per acre per annum. The farm is now staffed by a deputy farm manager, two assistants and six fishermen.

Tartabaw Fish Farm

This 40-acre farm, also managed by the PPFB, is situated 15 miles from the Twante Fish Farm and is fed by water from the River Irrawaddy. It is operated more or less in the same manner as the Twante Fish Farm. The salinity of water in the farm is less than that of Twante water.

6.3. Private Fish Ponds

The expert visited a good number of private fish ponds at Pawngde, Yandoon, Mandalay, Yamethin, Lawksawk, Nyangshwe and Kyawkse. Major carps are mainly cultured in lower Burma but in the dry zone, especially in and around Mandalay and Kyawkse, the culture of tilapia is more common. Most of the fish farmers were unable to furnish accurate data on the rate of stocking, quantity of artificial food given or the total production of fish per annum since proper records were not maintained, and in many cases, the information which was provided did not seem to be reliable. Consequently, no definite idea about actual fish production could be obtained. It was certain, however, that the majority of the farmers were getting satisfactory returns from their ponds. One of them stated that he was getting nine times more from one acre of pond than from one acre of land. It may be noted that the good returns to the fish farmers are in most cases not because of high production of fish, but because of the very high price of fish. There is ample scope for improvement in the culture practices followed.

The expert advised the fish farmers on scientific methods of fish farming, especially the utility of mixed farming, thinning out and harvesting of marketable fish, optimum rates of stocking with compatible species and related matters. In the case of tilapia culture, the advantage of proper manipulation of population by the regular harvesting of the larger fish and elimination of small fish (if there are too many) was explained. The majority of the fish culturists were observed to feed their fish with artificial food like rice bran and in a few cases with oil cakes. Rice bran is a comparatively cheap item of artificial food in Burma, but a number of fish farmers said they had difficulty in obtaining it in sufficient quantities. Most of the farmers depend entirely on the Directorate of. Fisheries for their supply of fish seed for stocking.

6.4. Future Development

Richness of the Resource

Burma is very rich in fishery resources. There is a large number of water areas, both large and small, suitable for fish culture, and a good number of important cultivable species. Among the indigenous ones are: *Catla catla* (Nga-thaing), *Labeo rohita* (Nga-myitchin), *L. nandina* (Nga-Ohn-done), *L. calbasu* (Mga-net-pya). and *Cirrhina mrigala* (Nga-gyin).

Besides, a few exotic fishes such as *Cyprinus carpio* (Hga-hpein, Israeli and Indonesian strains), *Tilapia mossambica* and *Osphronemus goramy* are also available for culture. Having such potentialities, fish culture in Burma could be greatly developed following proper planning and a well formulated program for long-range development throughout the country.

Present Government Plans for Promotion of Fish Culture

Projects planned by the Government of Burma for development of fish culture come under three main headings:

1. Survey of new sites for fish culture and establishment of model farms.

2. Production of an adequate quantity of fish seed for stocking.

3. Proper culture of indigenous and exotic fishes.

For extension of fish culture throughout the country, it is essential to explore new areas for construction of new ponds and the Government has taken up such a survey. As mentioned before, the Government has already established a few fish farms which are mainly used for the rearing of fish seed and its distribution to the public or for stocking their own farms.

At present stocking materials are collected from rivers, reared in nurseries in fish farms and then distributed to the public. The Government has already made extensive surveys of spawn collection sites. The quantity of spawn collected is increasing every year but the demand for fish seed is also increasing. The following table will give an idea of the increasing demands

Year	Total Spawn Collected from rivers	Total Number of Fingerlings Distributed
1961-62	600,000	70,000
1962-63	600,000	400,000
1963-64	900,000	600,000
1964-65	2,000,000	800,000
1965-66	5,000,000	1,000,000

The third important project is the proper culture of indigenous as well as exotic fishes. The Government has initiated culture of major carps on a large scale. At present a total of about 200 acres of ponds in Government farms are under cultivation and the area is going to be further extended to over 1,000 acres.

Common carp, *Cyprinus carpio* (Israeli strain) is also cultured along with the major carps to a certain extent, and the Government is contemplating the extension of common carp culture to higher altitudes. Accordingly, the expert accompanied by fee counterpart officer and the Director of Fisheries visited Maymyo in upper Burma and Taunggyi, in the southern Shan state. Both these towns are situated at an altitude of about 3,000-4,000 ft above sea level. The main purpose of the trip was to study the condition and advise on the feasibility of carp culture at such altitudes. At Haymyo town there are a few small impoundments in and around the botanical garden which might be taken up for the culture of *Cyprinus carpio* (Israeli strain) which thrives well in cold waters. Culture of the Israeli strain might also be introduced into the Shan states. A site near Shwenyang town 12 miles from Taunggyi, the headquarters of the southern Shan state, was tentatively selected during the visit as suitable for the construction of a fish farm for the distribution of seed of common carp for its propagation and culture in the area–if thought advisable.

The other exotic fish, *Tilapia mossambica*. first brought to Burma in 1952 released in the Hlawga Fish Farm ponds. The fish was bred and the fry were distributed to the public. But in spite of these attempts, Tilapia could not establish itself in lower Burma probably because of the abundance of predators like Wallago attu and *Ophiooephalus striatus* (Nga-yan). However, its culture seems to have developed to a great extent in upper and central Burma. Here it is cultured in small shallow ponds and even in cement cisterns.

In addition to these two exotic fishes, the Government is keen to introduce the Chinese grass carp (*Ctenopharyngodon idella*) in Burma as a biological control for submerged aquatic weeds.

Fish Culture in Lakes

Throughout Burma, there is a large number of small and big impoundments which could be used for fish culture. In Rangoon city itself there are two beautiful lakes suitable for fish culture. Moreover, Burma has two large fresh water lakes, Lake Inle and Lake Indiawgyi. A visit was made to Lake Inle in the southern Shan state. The lake, situated 3,000 ft above sea level, has a luxuriant growth of submerged aquatic vegetation comprised mostly of *Ceratophyllum, Hydrilla, Najas* and *Potamogeton*. The fish population in the lake includes very few commercially important species. The only economic species worth mentioning are *Ophiocephalus striatus, Clarius batrachus* and *Puntius sarana caudimarginatus*. There is also a variety of common carp endemic in the lake and adjoining areas called *Cyprinus carpio intha*. This variety is a very slow growing one. The fish production in the lake, as reported, is very meagre.

Tubb (see FAO/UN, 1957) suggested the introduction of the Chinese grass carp in the lake to control submerged vegetation. Accordingly, the Government of Burma made an attempt in 1957 to stock the lake with grass carp but without much success. This subject deserves further attention and an ecological survey preceding the establishment of suitable holding and rearing ponds to raise the introduced species to yearling size is necessary. Such a study may lead to a program of introduction of other species also.

Lake Indawgyi, in northern Burma, is the biggest lake. It has an indigenous population of major carps such as *Catla catla*, *Labeo rohita* and *Cirrhina mrigala*. Due to shortage of time and for security reasons, the expert was not able to visit the lake and give any suggestions for its development. However, he feels that many of these lakes and smaller impoundments could be utilized for fish culture which would ultimately help in increasing Burma's fish production.

6.5. Major Problems

Preliminary attempts made by the Government to develop interest in fish culture in Burma have been very encouraging. Private persons have become interested in the industry and taken up fish culture as a means of livelihood. The following table shows the increasing popularity of fish culture and gradual increase in the area under pond fish culture.

Year	Persons Engaged in Fish Culture	Total Pond Area (in acres) Under Cultivation
1961-62	50	100
1962-63	79	601.4
1963-64	171	1433.74
1964-65	252	1495.13
1965-66	378	1755-63

However, several major problems stand in the way of rapid development of the industry. The main problems are enumerated below:

A Need for Increased Seed Production

With the expansion of fish culture, the demand for stocking materials has also increased tremendously. Despite the Government establishment of spawn collection centers on the main rivers of Burma, results of wild fish seed collections from the rivers are not encouraging. The collections often contain a high percentage of non-economic varieties and the quantity is not adequate to meet the increasing demand. The alternative is to produce fish seed of the economic food fishes on a large scale through induced breeding of major carps by hormone injections.

Lack of Technical Knowledge

The lack of an adequate number of trained technical staff with knowledge of scientific pisciculture also stands in the path of rapid development of fish culture. To achieve success, an intimate knowledge of all the operations involved in nursery, rearing and stocking pond management is essential.

Need for Scientific Research

Very little scientific data are available in Burma on the limnological conditions of the fisheries waters and practically no research has been done on the problems of fish culture. Furthermore, there is a shortage of trained personnel to undertake such research.

The Government of Burma is making an earnest effort to solve the problems mentioned above. They have asked FAO for the services of a fish culture expert for a period of two years to advise and assist them on the development and extension of scientific fish culture, including surveys of areas available for fish culture, commercial production of major carp fish seed by induced breeding, initiation of selective breeding and hybridization, introduction and culture of suitable exotic species, and training of technical personnel.

The Government has also initiated research on problems related to both culture and capture fisheries and suitable staff to conduct research are being recruited. The Government has proposed the establishment of an institute for fisheries training, having a two-year diploma course. The institute is expected to start by 1970, and staff are being trained. At present, three students are undergoing training in the U.S.S.R. in marine fisheries and it is proposed that two more be sent to the Central Institute for Fisheries Education in Bombay, India, for two years of training in both inland and marine fisheries.

7.   General Conclusions

7.1. Expansion of Cultivable Areas

Fish culture in Burma is a new enterprise which is gaining popularity. Although scattered water areas could be made available for fish culture, the number of privately owned ponds is very limited.

Further expansion of fish culture requires the construction of new ponds, but in view of the Burmese regulation forbidding construction of ponds on land suitable for paddy culture, one has to seek out land areas which cannot be used for paddy. The Government has included in its program the survey of cultivable areas available for fish culture. In addition, it may be necessary to reclaim swamps especially those which are situated inside cities and towns. The moats in the towns of Handalay, Toungoo and Pegu could be reclaimed, cleared of weeds and used for fish culture. It is also encouraging that many private individuals are utilizing any little space available on their lands for construction of small ponds for culture of tilapia or major carps. The Directorate of Fisheries may provide necessary advice to fish farmers on this matter.

7.2. Procurement of Stocking Materials

The percentage of economically valuable species in the catches from the majority of the fry centers does not appear to be very high. Moreover, the total quantity of spawn collected is not adequate to meet the increasing demand for fish seed. Further surveys are necessary to establish new collection centers which will provide a better and larger supply of carp spawn. When better centers are established, those producing small quantities of major carps may be abandoned.

It is also essential for the Government to develop the technique of induced breeding of carps by hormone injections. In fact, successful commercial production of pure fish seed should be able to solve the fish seed problem. It is hoped that with the help of the next fish culture expert, who will assist the Government for two years, enough local officers will be trained to be able to produce millions of fish seed for stocking all inland waters.

In addition, the possibilities of constructing 'Bundh' type of ponds for carp breeding could be explored. The topography of the undulating terrain about 38 miles from Rangoon on the Pegu Road and amount of rainfall and catchment area could be studied. If found suitable, an attempt may be made to construct one such pond for experimental carp breeding.

7.3. Establishment of Composite Fish Farms

With an enlarged fish culture program it will be necessary to have a composite fish farm self-sufficient in its requirement for fish seed. This farm should have the facilities for induced fish breeding experiments, and a set of nursery, rearing and stocking ponds in addition to marketing ponds and ponds for rearing breeders.

The Government of Burma has two production fish farms under the management of the People's Pearl and Fisheries Board. One of these, the 104-acre Twante Fish Farm, will be extended to 1,000 acres by 1970. This huge farm has a proposal for large-size (mostly 18 acres) stocking ponds and a few marketing ponds but no provision for nursery and rearing ponds. The Board depends on the Directorate of Fisheries for the supply of carp fingerlings. Even now, it is difficult to supply the quantity of proper size fingerlings required for stocking and when the farm grows to a size of 1,000 acres, the requirement of fingerlings for the farm alone would be 3 million, calculated at the stocking rate of 3,000 per acre.

The expert has suggested therefore the construction of an adequate number of nursery and rearing ponds for raising sufficient carp fingerlings for stocking 1,000 acres. As a preliminary project, it would be worthwhile to have one set of nursery and rearing ponds, sufficient to provide stocking materials for 50 acres of stocking pond. A rough calculation of production and survival rates at various stages was made and it was suggested to have one acre of nursery area and four acres of rearing pond for every 50 acres of stocking ponds. One acre of nursery ponds could also be utilized for rearing purposes. After studying the results of production and survival of spawn and fry at various operational stages, a program could be set up for the extension of more sets of nursery and rearing ponds. In order to become self-sufficient in fish seed requirements, the farm should start its own induced fish breeding program and grow enough breeders for the purpose. It would also be necessary to convert a few smaller stocking ponds for maintaining a stock of breeders.

Other small farms could also be made into composite fish farms and used as demonstration areas for the training of the fishery staff, extension workers and fish farmers.

7.4. Culture of Major Carps and Other Indigenous Species

In Burma, the major carp *Cirrhina mrigala* (Nga-gyin) is the favourite fish and is sold at the highest market price compared to other carps. So there is a tendency to stock more mrigals which often results in their poor growth and less production. The necessity of stocking compatible species in definite proportions should be stressed. Production can be increased by mixed farming, that is stocking with a number of species of fish having different feeding habits.

One of the indigenous carps, *Osteobrama* (Rohtee) *belangeri* grows to a fairly big size and seems to be a suitable fish for culture. However, before culturing this species along with major carps, it would be advisable to study its biology, food and feeding habits and rate of growth.

At present, fertilization of ponds is not regularly practised, and it is difficult to suggest any definite methods of fertilization without knowing the quality of the soil and other ecological factors. In the future a program for pond fertilization should be taken up.

Fish farmers in Burma almost invariably feed their fish with artificial food, e.g., rice bran, and it is important to develop a supply of economical and effective artificial food.

There is ample scope for improvement of nursery management practices for obtaining higher production and maximum survival of fry.

7.5. Culture of Exotic Fishes

Common Carp

At present, Burma has three different strains of common carp (*Cyprinus Carpio*), one is indigenous (var. intha), one was imported from Indonesia and one from Israel. Though these strains were introduced several years ago, they have not become very popular.

It is doubtful how well the more domesticated Indonesian strain could thrive in wild waters which are full of predatory fishes. The Israeli strain should prove much better, especially for culture in cold waters at the higher altitudes. However, it should be noted that Burma already possesses a good supply of major carps and concentration on successful culture of these appears to be more advisable than turning attention to *Cyprinus carpio* at this time.

Tilapia Culture

Tilapia mossambioa, though introduced in 1952, do not seem to have established themselves in ponds and other waters, especially in lower Burma, probably because of the presence of predatory species. The fish could be cultured profitably in the dry zone area of upper and central Burma in small domestic ponds and also in cisterns. Fish farmers who are culturing them are reaping quite a good harvest. Besides, the greatest advantage of culturing tilapia in Burma is that even the smallest size fishes would be utilized for the preparation of the fish sauce, Ngapi. Proper culture of this species should be encouraged. The Government should consider establishing tilapia seed distribution centers in upper and central Burma for the distribution of seed to the public.

Tilapia-cum-Murrel Culture

The most common murrel in Burma, *Ophiooephalus striatus* (Ngayan), is well liked by the people. It is, therefore, suggested that a program for tilapia-cum-murrel culture might be initiated. Proper manipulation of stocking rate and number should provide good production of suitably sized tilapia as well as murrels.

Culture of Chinese Carps

The Chinese grass carp *Ctenopharyngodon idella* has great demand because of its ability to control submerged weeds. A consignment of this species was brought by the Government for stocking Inle Lake during 1957. A few of the adult fishes are still being held at the Hlawga Fish Farm. Its introduction is most likely to be a welcome addition to the lot of indigenous species. In case the Government decides to initiate a program of systematic stocking and culture of this species in Inle Lake, the proper technique of rearing and breeding the fish should be developed. The fish culture expert would be in a position to train local workers in the technique of rearing and breeding of the species.

The Government may also explore the possibility of introducing the fast-growing Chinese silver carp, *Hypophthalmiohthys molitrix*. A small consignment may be cultured at the Hlawga Fish Farm to study its effect on other species before introducing it in other waters.

Paddy-cum-fish Culture

The culture of tilapia in the rice fields of Burma was initiated in the year 1955. According to FAO/UN (1957) the preliminary results were rather encouraging but the project was dropped in 1957 because of unsatisfactory results probably due to destruction of the tilapia by predatory fish. Before rejecting the project forever it may be useful to determine the reasons for unsatisfactory results and proceed on the basis of the findings.

7.6. Training of Personnel

At present, there is a dearth of trained personnel. The Directorate of Fisheries has expanded its fish culture section and it is essential that newly recruited personnel should receive proper training in scientific fish culture and fish farm management. The FAO fishery expert who is expected to assist and advise the Government of Burma during 1967-68 would be able, during his period of assignment, to train as many personnel as are available. In addition, it may be necessary to send some staff abroad for further training.

Some of the State agencies also seem to be understaffed, especially the Twante Fish Farm. With further expansion of its area and its technical activities, it will be necessary to increase the staff considerably.

7.7. Summary of Recommendations

Survey and Reclamation

A proper survey should be made of potentially cultivable waters and the reclamation of suitable areas should be undertaken to extend fish culture to fresh areas.

Production of Fish Seed

Better spawn collection centers should be established and the commercial production of fish seed of major carps through injection of pituitary hormones should be taken up in earnest.

The expert has submitted a list of the equipment and material required for fish breeding experiments to the Director of Fisheries. This equipment should be obtained before the ensuing fish breeding season so that the fish culture expert can assist in the production of fish seed and training of personnel at that time. The list also includes the total requirement for pituitary glands, total number of breeders and field staff required for the production of ten million quality fish seed.

Composite Fish Farm

Twante Fish Farm should be converted into a composite fish farm, self-sufficient in its requirement for stocking materials. The farm should have an adequate number of nursery and rearing ponds for rearing fry for the stocking ponds and facilities should be provided for large-scale production of carp seed for its own requirement.

Culture of Major Carps

Steps should be taken for further improvement of nursery management and other fish cultural techniques to ensure a higher rate of survival and production.

Culture of Exotic Species

(a) Some attention might be given to the cultivation of common carp (the Israeli strain) at higher altitudes but not at the expense of successful development of the indigenous major carps.

(b) Tilapia culture should be popularized by establishing more seed distribution centers, especially in upper Burma, and supplying seed to the public. Tilapia-cum-murrel culture should be tried.

(c) Further attempts should be made to introduce grass carp in Inle Lake to control aquatic weeds.

(d) The usefulness of introducing silver carp might be explored.

(e) The reasons for unsatisfactory results from paddy-cum-fish culture with tilapia might be investigated.

Training

Two of the technical staff should be granted fellowships for one-year training courses in the Central Inland Fisheries Research Institute, India. One of them should have an adequate knowledge of chemistry so that he can obtain training in water and soil chemistry.

One officer with experience in fish culture in Burma should be sent to a foreign institute where training facilities in the management of impounded waters, forage-cum-pre-datory fish culture and pond fertilization are available.

References

Ling, S.W., 1955. Report to the Government of Burma on inland fisheries development. Rep. PAO/EPTA, 361: 9.

FAO/UN, 1957. Report to the Government of Burma on inland fisheries development. Based on the work of J. Alan Tubb, FAO Expert in Inland Fisheries. Rep. FAO/EPTA, (776): ii + 64 p.

(D) Grass Carp Project in Fiji

It was very short duration project two weeks from June, 2 1974. Prof. H. Chaudhuri was then holding the post of Head, Fish Culture Division and Officer-in-Charge of Central Fisheries Research Sub Station, Cuttack.

Relevant extracts from Prof. Chaudhuri's report is reproduced with some modifications to impress how his ingenuity, practical field experience, prompt conception of problems and research technology solve the intricate problem in a short visit that saved the state from huge expenditure and opened up ways to enhance protein produciton.

1.   Terms of Reference

To study the present status of the Grass Carp Project including study of the field conditions and facilities available for rearing and breeding of grass carp and to assist and advise the Government of Fiji on the proper rearing of the grass carp fingerlings recently sent to Fiji for building up of a stock of grass carp for its propagation and their subsequent release in the rivers for the control of submerged weeds.

2.   Programme Development

The expert was in close touch with the office of the High Commissioner for India in Fiji. At the instruction of the High Commissioner he prepared an interim report and submitted it for discussion with the Secretary, Ministry of Agriculture, Fisheries and Forests so as to facilitate proper implementation of the Grass Carp Project. On the basis of the report discussions were held at a meeting on June 13, 1974. Mr. Winston Thompson, Secretary (Agriculture) and Mr. Goswami, Deputy High Commissioner for India in Fiji were present in the meeting in addition to the officers of Fisheries Division.

The expert was invited to deliver a lecture at the University Zoology Department on "Fish Culture and Fish Breeding".

3. Origin of the Project: Past Studies and Present Status

The Department of Agriculture carried out investigations since 1960 on the problem of weed infestations in Rewa river system. Detailed studies were made on the location, soil and climatic conditions of the catchment area of the Rewa river. Mrs. H.R. Hughes, Botanical consultant, Department of Agriculture, in her report entitled. "The ecology and biology of the Rewa river system, Fiji with particular reference to aquatic weeds" has given on account of the growth of weeds in the Rewa river since 1960 to the beginning of 1969 and also described the dredging operations and actual river use by boat traffic. She has studied the area of infestations and distribution of the weeds in the rivers and discussed all possible means of control of these weeds including the use of mechanical weedcutters and chemical weedicides and recommended the use of grass carp as the most effective method of controlling the weeds. According to her, "serious weed infestation covers 600 ha and that 30 grown up carp (over 1 kg) are necessary to control weed in 1 ha, a carp population of 18,000 would have to be built up in the Rewa". Fisheries Division on the other hand calculated a much higher figure by their preliminary study made by introducing grass carp in a pen in the river. According to them, "125 carps/ha or 1,500,000 grass carp are required to control the weeds". These figures, maybe considered as very tentative since the estimation was not based on adequate scientific observations. The studies made, however, led to the conclusion that the menace of weed infestation in the rivers could only be effectively and economically controlled by the biological control agent grass carp and the Fisheries Division decided to import grass carp for the purpose.

The first consignment of grass carp brought to Fiji was as early as in December, 1968. 200 grass carp fingerlings along with 60 silver carp and 80 big head were imported from Malacca Research Station, Malaysia. They were reared in a fish pond (0.2 ha) at the Lami Fish Farm. In September, 1969, 30 (1 kg each) grass carp were released in a pen with heavy infestations of *Hydrilla* in the Rewa river near Nausori bridge for studying the capacity of these fishes to control the weeds. But unfortunately the study was vitiated and terminated due to sudden flooding of the river when all the experimental fishes were lost. The remaining stock of grass carp was reared in the pond and by the end of 1972 only 30 adults (6.8 kg average) were surviving. These fishes were fully mature. But the entire stock died due to lack of space in the holding pond and consequent depletion of oxygen.

During 1972, a second consignment of 1000 fingerlings of each of the three species of grass carp, silver carp and bighead were imported from Taiwan but all died during the transport. A third lot of identical numbers were again brought from Taiwan at the beginning of 1973, 60 per cent of which were found dead when received. The rest of the stock were reared at the Lami Fish Farm, for a short period and then released in the Tonga River in March, 1974.

The last consignment of 2,000 fingerlings of grass carp were received from the Central Inland Fisheries Research Sub-Station, Cuttack, India, on March 24, 1974. Only 25 fingerlings died during transport and the rest were being reared in a cement cistern at the Lami Fisheries Station. About 10 or so were reported to have died subsequently by jumping out of the cistern. The fingerlings appeared healthy and were being fed with *Hydrilla* collected from the river and also with supplementary feed copra meal and rice bran.

In addition, the Fisheries Division, Fiji also imported *Puntius gonionotus* from Malaysia especially for the purpose of using the species as donors for induced breeding of grass carp. Nearly 95 per cent of the fry were reported to have died during transport. The rest were being reared in one.057 ha (1/7 acre) pond at the Lami Fish Farm. A small stock had been built up due to their natural spawning during the last two years.

Lack of pond space was reported to be the main difficulty experienced in raising adequate numbers of the imported species to suitable size for their propagation. The pond area available at the Lami Fish Farm is limited to one 0.2 ha (1/2 acre–Pond No. 3), one 0.057 ha (1/7 acre–Pond No. 2), one 0.04 ha (1/10 acre-Pond No.1) and two 0.02 ha (1/20 acre each–Pond Nos. 4 and 5) ponds. Pond 2 was stocked with *Puntius gonionotus*. Pond 1 was reserved for *Tilapia nilotica* (proposed to be imported from Israel) and Ponds 4 and 5 were under repair. Pond 3 was infested with Tilapia and was drained and being refilled.

4.   Weed Infestation in Rivers

Preliminary observations were made for ascertaining the degree of infestations of the weeds in the rivers. Infestation by *Hydrilla* was very little in the Waimanur River. It was reported that due to cyclone and flood large quantities of the weeds had been washed off. In the Rewa River, it was observed that the marginal fringes had dense growth of *Hydrilla* at several locations. *Potamogeton* was seen further down. The Tonga River especially near its origin from the Rewa River had deep infestation of *Hydrilla*. It could not be however possible for the expert to make quantitative evaluation of weed infestations throughout the length of Rewa River system during the short period of his stay at Fiji.

4.1. Fish Population

Stray observations were made on the fish population of the Rewa river system. Fishing was done by cast netting and seining at certain locations in the Waimanu, Rewa and Tonga River. Hardly any indigenous freshwater fish of commercial importance was encountered in the rivers. *Tilapia mossambica* introduced in the rivers of Fiji during mid-fifties, was encountered in several catches from different locations of the rivers. Besides, a number of brackishwater fishes, such as gobies, mullets, *Caranx* sp. and the glass-perch (*Kuhlia rupestris*) were caught in the tidal zone. The glassperch locally called iko draka is found in most of the rivers. It is reported that the tidal zone extends as far as 40 km from the Rewa river mouth and that predators like sharks, *Caranx* sp. and the rupestris migrate even up to 50-65 km up. Freshwater eel, a minor predator is also present in the river.

Mrs. Hughes have mentioned in her report of several species of shell-fish from the river. Commonly among those are *Batissa violacea* (Kaibuli and Kai Jamu), *Unio* sp. (Kai ni waid-ranu), *Melania* sp. (motomoto) and *Neritina* sp. (yamicoko and kadrudru). The mussels are very common and are eaten by the Fijians. Quite a few of these mussels were collected from the river bed.

The only commercial species of prawn available in the river is *Macrobrachium* larvae which sometimes get inadvertently introduced in the ponds being brought along with the weeds. This species can be cultured in ponds.

The professor then studied the issue of feasibility of introduction of grass carp fry or fingerlings into the river, facilities available for rearing of grass carp fry and fingerlings, sites where fresh fish ponds could be established and very other assciated issues are made the following recommendations.

4.2. Rearing of Grass Carp Fingerlings

With a view to solving the weed problem of the Rewa river system by introduction of grass carp, great care has to be taken to rear the present stock of fingerlings imported recently from India so that the stock may form the nucleus from which large-scale production of grass carp could be made for stocking the rivers from time to time. It would be essential to have a full-time Fisheries Officer for taking proper care of the present stock. The officer with the occasional help of two to three supporting staff should be able to carry on the work for the present.

The stock of grass carp in the cement cistern had grown to two distinct size-groups. For the normal growth of the smaller fingerlings it would be necessary to segregate the bigger ones and transfer those to Pond 3 after carefully refilling the same with water. Healthy *Hydrilla* should be supplied regularly in excess of their requirement taking care that water does not get polluted by the rotting of leaves. In addition, supplementary feed such as oilcakes, copra meal, wheat short or rice bran may be fed. Care should be taken not to introduce excess of these feeds and supplementary feeding should be stopped in case the water develops a thick algal bloom. Routine analyses of pH and dissolved oxygen content of water may be made every fortnight. To prevent appearance of thick algal bloom and consequent depletion of oxygen, it is suggested to have an additional constant flow of spring water (as proposed) through a pipe into the Pond 3. The water may be let in, in the form of a spray so that the pond gets properly oxygenated. Plankton sample may be collected fortnightly and examined. Records of various data including growth rate, quantity of feed given etc. may be regularly maintained.

The smaller fingerlings in the cistern may be fed with cut bits of healthy *Hydrilla* and also with powdered artificial feed. These fingerlings may be transfered from the cistern to the pond when they attain about 100-125mm size. Since it would be risky to hold all the growing fish in one 0.2 ha pond (Pond 3), it is essential that two more 0.2 ha ponds be constructed at an early date so that the fishes could be distributed to the three ponds for their proper growth and health.

After proper survey of the Naduruloulou first site is to the nutrient level of the soil, its retentivity and also availaibility of adequate water from the adjacent stream, initially two 0.2 ha ponds each having a size of about 100 m x 20 m and a depth of about 1.6 m may be constructed for the rearing of the fingerlings to adult fish.

4.3. Rearing of Breeders for Propagation of Grass Carp

When the fingerlings attain the size of 250mm or above, they may be restocked and redistributed in the 3 ponds (total area 0.6 ha) at the rate of 1,000 fish per ha for raising those to probable breeder for carrying out induced breeding experiments. Thus having assured of a good stock of 600 fish for raising breeders, the rest of the stock (over 250mm size) may be released in the purely freshwater zone in the upper region of the Rewa river and its tributaries. However, before release of the yearlings investigation may be made as to the weed infestations, depth of water, water quality and the fish fauna of the area.

The fishes restocked in the ponds may be regularly fed with sufficient quantity of *Hydrilla* supplemented by small quantity of artificial feed. The *Hydrilla* collected from the river may be thoroughly washed and kept for about a day before introduction in the pond. This would prevent entry of Tilapia and any other weed fishes in the pond. A rectangular frame of bamboo or wood may be made in the pond inside which the aquatic weeds may be introduced for feeding.

If properly raised the present stock of grass carp may mature fully by December, 1975 when the fishes would be about 2-1/2 years old. In the tropical waters of India, grass carp male matures even in one year but females take at least two years to mature. At a lower temperature in Fiji it is not certain whether the fishes would mature at this age but it is expected that at least a small percentage of the stock of breeders might mature by the end of 1975.

Keeping this aspect in view, raising of as many as 600 probable breeders has been recommended. By the end of 1975, the breeders are expected to attain about 2.0 to 2.5 kg in weight and in the event of their attaining full maturity attempts have to be made to breed them by injections of fish pituitary hormones.

4.4. Raising of Donors for Pituitary Glands

Before carrying out any experiments on induced breeding of grass carp it is essential to procure an adequate stock of pituitary gland materials for injecting the breeders. *Puntius gonionotus* was imported from Malaysia with the object of using it as a donor fish for collection of pituitary glands. It is a suitable donor fish and can be reared and bred without much effort. But some pond space have to be provided for breeding and rearing of these donor fishes. About 500-600 specimens of mature *Puntius gonionotus* may have to be raised so as to ensure sufficient stock of pituitary glands. *P. gonionotus* may be reared in smaller ponds initially and then release in the impoundment at the Naduruloulou Research Station as soon as the latter is constructed. As in the case of grass carp, aquatic weeds as well as supplementary feed may be provided for rearing the species. For breeding, fully mature male and female breeders may be selected and released in smaller ponds which had been previously drained out and freshly refilled, preferably with a constant flow of water.

Mature fishes should be selected for gland collection. The pituitary gland, situated vental to the brain proper may be carefully removed by dissecting the head, washed in absolute alcohol and finally preserved in pure absolute alcohol. At the time of injection the glands may be taken out of absolute alcohol, weighed quickly in a chemical balance and the required quantity macerated in distilled water for the preparation of gland extract.

4.5. Preparation for Induced Breeding of Grass Carp

As mentioned in the present stock of grass carp if properly raised, may attain full maturity by the end of 1975 when they are 2-1/2 years old. It was observed that a few grass carp brought from Malacca in December 1968 matured in December 1971 when they were little over three years old. It is expected that a few grass carp of the present stock would attain full maturity by December 1975 or at the beginning of 1976.

To initiate a programme on induced breeding some preparatory arrangements have to be made. For large-scale production of grass carp seed and rearing of large number of fry and fingerlings it would be necessary to construct a few more ponds. For this purpose, the barest minimum would be to construct two 0.04 ha nursery ponds, two 0.1 ha rearing ponds and two 0.4 ha stocking ponds. Besides, facilities for obtaining fresh rainwater, if made available, would ensure higher success in the induced breeding programme. Action may be taken to make these ponds available before the breeding season i.e. December 1975. These ponds may be constructed at the Naduruloulou Research Station adjacent to the site where two 0.2 ha ponds have been proposed to be constructed. The availability of adequate water for these ponds from the adjoining stream may be ascertained before construction of the ponds.

For conducting induced breeding experiments it may be necessary to get the assistance of a grass carp breeding expert. Assignment of a grass carp expert with the provision of proper field facilities for work and raising of an adequate stock of breeders, the service of a grass carp expert may be sought for a period of 1-1/2 years. The expert may arrive a couple of months prior to the breeding season i.e. December 1975 for making preliminary arrangements for grass carp breeding. His services thus can be utilized for two breeding seasons.

His terms of reference would be to advise and assist the Government of Fiji in their Grass Carp Project for the effective control of the submerged weeds in the Rewa river system by release of grass carp with special attention to the breeding of grass carp in ponds by administration of fish pituitary hormones, hatchery and nursery management, rearing of fingerlings and raising of breeders; training of local workers and related duties.

5.   Conclusion

In conclusion, it may be mentioned that the Fisheries Division, Ministry of Agriculture, Fisheries and Forests have rightly indicated that the problem of weed infestations in the Rewa river system could be effectively solved by grass carp, the biological control agent. Importing of large number of grass carp fry from foreign countries and releasing those directly into the river is not only a difficult task but is likely to prove very expensive and at the same time may prove ineffective since the river harbours a good population of highly predatory fishes. The only alternative to this important problem is to raise the present stock of grass carp to breeders, breed them successfully and build up a huge stock of grass carp for their systematic release in the river. This no doubt involves quite a bit of expenditure initially for construction of new ponds, but once constructed the same ponds can be suitably utilised for production of fish for food.

In future, a self-contained small fish farm complex of an economically viable 4 ha water area may be planned having six 0.4 ha, four 0.2 ha, four 0.1 ha and four 0.04 ha ponds and some space for a hatchery. By the end of 1975 eight of these ponds would be ready for the grass carp Project. The rest of the construction work may be done on phased basis covering a number of years for its completion. The blue print of the farm may be prepared now but the construction work be taken up as and when funds are available. A good, experienced engineer's service may be sought for the construction. This farm would initially be utilised for producing grass carp for control of weeds in the Rewa river system. Finally, it would form a base and a training centre for the development of aquaculture in Fiji. Polyculture of several quick growing food fishes like the Indian and Chinese carps would give high production of fish in ponds and provide valuable protein to the people of Fiji. These important riverine species when released in the rivers would enrich the riverine fish fauna of the islands. The grass carp would not only control the submerged weeds in the rivers but it would provide thousands of kg of fish by converting the weeds into fish flesh. In addition, the semi-digested fecal matters of grass carp would enrich the water and provide food for other fishes especially the bottom-dwelling ones and also to the shell-fishes thus increasing the production of mussels providing additional food to the Fijians. The expert is convinced that if the grass carp project is continued as suggested, the money that would be spent for the construction of ponds would be realised in no time once success in breeding grass carp is achieved and grass carp successfully released in the river system.

While concluding it may be worthwhile to quote the remarks made by Mrs. Hughes about grass carp in her report "Not only may it be possible to control definite weed areas quickly hut a secondary source of income and food may result".

(E) Short Duration Malaysian Carp Project (1978 March 20th to 30th): Professor Chaudhuri's 1st Consultancy was in 1975 Jan-Feb

Introduction and Background Information

The culture of Chinese carps is very popular in Malaysia. In the absence of an effective technique for obtaining seeds of these quick growing cultivated species, the Malaysian fish farmers were solely dependent on the import of millions of carp fry every year, involving considerable drainage of valuable foreign exchange of the country. With a view to help solving the major constraint of seed supply by developing a standardized technique of mass production of seed of Chinese carps by induced breeding, the International Development Research Centre (IDHC), Canada sponsored a research project in collaboration with the Malaysian Agricultural Research and Development Institute (MARDI), primarily at the Freshwater Fisheries Research Station at Batu Berendam, Malacca in October 1974. The duration of the carps (Malaysia) project was initially for three years. On completion of the Phase I of the Project the same has been extended to a further period of three years. In the Phase II of the project attention has especially been given to develop standard procedures for induced spawning of several important indigenous food fishes, in addition to the continuation of research on further improvement on the breeding technique of Chinese grass carp.

Dr. H. Chaudhuri, Regional Aquaculture Coordinator, IDRC/ SEAFDEC Milkfish Project, visited the Freshwater Fisheries Research Station at Batu Berendam, Malacca for making an assessment of the completed research programmes of Phase I of the Project and also for assisting and advising the Project Leader on the technical aspects of the Phase II of the Project, especially on the methodology to be followed in various research programmes.

The consultancy visit of Dr. Chaudhuri lasted for a short period of 11 days from March 20 to 30, 1978.

The consultant is thankful to Mr. Ahmad Tajuddin, Project Leader and Head, Freshwater Fisheries Station and his colleagues for their generous help, rendered in his work both in the field as well as in the laboratory lie had some useful discussions with Dr. Eric Watts, Project Advisor (Nutrition) although Dr. Watts was busy in preparing for his departure. The consultant is grateful to both IDRC and SEAFDEC Aquaculture Department for arranging the visit.

1. Phase I of the Project

Objectives

1. To develop standard procedures for bulk production of fish seed, at all times of the year, by induced breeding of carps using salmon gonadotropin;

2. To develop a pilot hatchery for the production of seed fry and train personnel in fish reproduction, nutrition and hatchery techniques for its practical operation;

3. To investigate local sources of gonadotropin from tuna or other fish;

4. To determine the nutritional requirements of the various stages of the fish being investigated; and

5. To develop suitable feeds from a locally available and inexpensive materials.

2. Programme of Research

Induced Spawning

At the initial stage the project did not make satisfactory progress mainly because of lack of experience of the young research staff of the station in the technique of fish breeding by neuro-hormonal treatment. The consultant visited Malacca station previously during Jan-Feb 1975 as an IDRC consultant, when he demonstrated the technique of pituitary injection and gave instructions for the proper maintenance of the; brood stock and on all other aspects of induced breeding including hatchery techniques. He made several recommendations and also suggested training of the counterpart, officer in induced breeding of Chinese carps and nursery management at the Central Inland Fisheries Research Station, Cuttack, India. Accordingly, far Tajuddin (the present Project Leader and Head, FFRS) was sent to Cuttack training for about 6 weeks in 1975. This was followed by the visit of another researcher Mr. Haron to Cuttack, India and Nepal for similar training. The training of these two MARDI officers had greatly helped in the successful implementation of the spawning programme and the first successful spawning in bighead with viable eggs was obtained soon after the return of Mr. Tajuddin from his training.

Although use of salmon gonodotropin was envisaged in the objectives of the project, the spawning team's experiments with SG-G100 were not very encouraging as compared to those with carp pituitary gland extracts, SG-G100 needed the same dose as the carp pituitary for successful spawning, but more consistent results were obtained in the spawning of both bighead and silver carp with carp pituitary glands. Other salmon preparations such CON A (14), Extract 13, Extract 13+ and crude salmon pituitaries were also tried on test fishes.

The first two hormones needed 33 times higher doses for the precipitation of spawning as compared to carp pituitaries. The dose, appears to be too high. It is not known whether the experiments were conducted simultaneously under similar environmental conditions and keeping controls. It might also be possible that the effectiveness of the hormones deteriorated due to storage. Subsequent experiments using HCG as the primary dose followed by a low dose of carp pituitary gland gave highly successful results and the Project staff was able to breed silver carps and bigheads almost the year round or obtaining, their seed. With grass carp, however, the result was less successful. Techniques followed by various institutions were tried and successful results were obtained on several occasions. The technique, however, needs further refinement and standardization. Proper maintenance of large number of brood stock and large number of trials under varying climatic and environmental conditions would help in standardization of the technique.

Hatching Technique

The use of hatchery jars (156 L capacity), made of fiberglass was instrumental in improving the hatching rate. Occasionally premature hatching was observed due to the use of rainwater (soft water) which was greatly remedied by treating tin eggs with 0.52 g Tamvic acid per litre. The acid is administered to the hatching jar about 8 hours after fertilization of the eggs. This, however, delays the incubation period from the normal 15-18 hours to 24-30 hrs. This study needs further standardization since delay in hatching (almost double the normal incubation period) involves more risk and more expenditure.

Nursery Management

Nursery management technique has been greatly improved. Trials with various organic manures were given and ground soyaben applied at a rate of 3 kg/10 aq.m was found to be the best. This manure produces heavy swarms of microc rustacens (Moina). Intermittent low doses of urea may help in maintaining the peak.

The stocking rate of post-larvae used (3.75m/ha) is quite high. For such an intensive stocking rate the natural feed should be supplemented by artificial feed. Both the growth (2.6-3 cm in one month and survival (30-40 per cent) can be improved by supplementary feeding and stock manipulation (thinning out). The control of aquatic weeds and filamentous algae should not be a very difficult problem in such small nurseries (0.004ha). These can be eradicated manually. The control of aquatic insects (notonectids) is being done by spraying soap-oil emulsion, a method demonstrated by the consultant during its first consultancy visit.

Nutrition of Grass Carps

Very little progress has been made in Phase I of the project on this important aspect of research, especially on the determination of nutritional requirements of various stages of grass carp and formulation of suitable feed from locally available cheap materials. The reason put forward for this was that the study could not be initiated early due to late arrival of equipment and delay in installation of the nutrition lab.

Live Feed Culture

Preliminary efforts made in producing live feed such as Brachionus, Moina and Daphnia were very encouraging. The best feed for post-larvae was found to be live feed. The larvae should, however, be fed with minute live feed within 48 hours after hatching, when the hatchlings start a directive movement. At this stage the mouth is fully formed and the yolk is largely absorbed. Delay in feeding may lead to high mortality.

Diet Formulations

The consultant did not have the opportunity to go through the final report of Dr. Eric Watts, Project Adviser (Nutrition) to learn about the work conducted by him on the nutritional aspect of the project. However, Dr.Watts kindly gave him a copy of the report on

"Determination of the Nutrient requirements of the grass carp" prepared by Mr. Allan Castledine, a CIDA scholar. He conducted two preliminary experiments on the (1) determination of an optimum dietary protein level, and (2) determination of optimum protein, level and optimum protein to energy ratio for juvenile grass carp. The results of the two-week experiment (first experiment) showed that maximum weight gain was indicated at 40 per cent dietary protein beyond which the growth was depressed. The conclusion made by the worker that "grass carp may be more efficient than other cultivated fish species in utilizing low levels of dietary protein is rather hasty and not very justified since conversion value is less than half of that of 40 per cent and in ad libitum feeding the fish may get substantial quantity of protein for growth since less concentrated diet can meet the requirement if large quantities of it is provided". In the second experiment too, the conclusion arrived at does not appear to have been based on the actual result since the diet containing 26 per cent protein with an M.E. of 3.46 k cal/g appeared to be the best.

Pituitary Bank Fractionation of Tuna Pituitaries

The study (grant-in-aid project) undertaken by Dr. Khoo Khay Huat at the Universiti Sains Malaysia, Penang on testing pituitary glands of tunas obtained from the canneries in Penang and comparing that with salmon gonadotropin did not meet with much success because of poor quality of donor fish. Due to the marketing policies of the company good quality fish were frozen as whole-fish and collection of pituitary gland was not allowed. The limited experimental study on testing the crude tuna pituitary extract showed very negligible amount of gonadotropin. As reported by Dr. Huat, attempts to develop simple bioassay on juvenile Trichogaster pectoralis also was not successful.

3. Phase II of the Project

Objectives

1. To develop standard procedures for induced breeding of some of the local fish of importance, such as *Leptobarbus hoevenii* (Jelawat), *Tor tambroides* (Kelah), *Puntius bulu* and also Chinese grass carp. *Ctenopharyngodon idella* (Kam rumput);

2. To develop methods by means of the pilot hatchery for the production, nutrition and transport of the species (such

as silver carp and bighead carp) for which fry can now be produced by induced breeding and other species for which breeding is later achieved;

3. To continue the investigations of Phase I of the project after determining the nutritional requirements for the various stages of the fish being investigated and additional fish of local importance;

4. To improve the existing methods of fry culture, including nursery and rearing pond management for the fish being investigated;

5. To make preliminary investigations on methods of stocking in selected natural freshwater bodies and also artificial impoundments in Malaysia;

6. To continue investigations for local gonadotropin sources, including common carp, *Puntius gonionotus* and *tuna*; and

7. To transfer the improved technologies related to induced breeding, nursery and rearing pond management and transportation of fish fry and fingerlings to the Federal Fisheries Department.

4.   Programme of Research

The consultant had detailed discussion with Mr. Taj Uddin, Project Leader and Head, FFRC on the research programmes to be formulated based on the above objectives, priorities to be fixed up methodology to be followed. The consultant gave his views and suggestions on relevant matters and a tentative schedule activities prepared.

Seed Production by Induced Breeding

Indigenous Species

In Phase II of the project more stress has been given to develop standard procedures for induced breeding and seed production of a few important indigenous commercial species of carps. Regarding the biology, food and feeding habits, breeding habits and cultural aspects of these local species of carps very little information is available. So before making any suggestion on the programme of research to be taken up on these species the consultant decided to consult literature to collect all available information on these species.

Tor tambroides (Kelah)

The species is available in Indonesia, Thailand (Petchaburi R.) and in Sarawak, East Malaysia (upper Sardong R.). It grows up to 90 cm and 21 kg in weight. 50 cm maximum size is recorded in Thailand. Its supply is very limited. The fish is considered "best among freshwater fishes in Sarawak where successful culture has been carried out in ponds with good flow of water and in the presence of stones and rocks". Practically nothing is known about its feeding and breeding habits. Like Indian Mahseer, it is adapted to rocky bottom in rapid flowing streams. Its growth in stagnant ponds is expected to be very slow.

Immediate attempts should be made to procure fry or fingerlings from Sarawak, carely transported them and stocked in ponds at the Research Station. The gut-contents of the various stages of the species caught from the natural habitat may be studies so as to get some idea of its food habits for facilitating proper rearing of the species. Suitable supplementary feed may be formulated and facilities, if available, for maintaining a mild flow of water in the pond may be provided.

Puntius bulu (megalan)

The species has distribution in Perak (West Malaysia), Saba and Sarawak (East Malaysi), rivers of Sumatra and Kalimantan (Borneo) and Lakes of peninsular Thailand. It grows up to 3.6 to 4.2 kg. The fish feeds on worms, plant materials, insects and detritus. It is reported to be suitable fish for pond culture but information is available as to its rate of growth in ponds and breeding habits.

Procurement of fry, fingerlings and adult fish, if available has to be made. As in case of Tor tambroides, a stock has to be built up by rearing the seed in fertilized ponds and provided with supplementary feed. The growth rate may be studied. In case spawners are available attempts have to be made to spawn them by providing proper facilities and also by hypophysation.

Leptobarbus hoevenii (Jelawat)

Jelawat is available in Indonesia (Sumatra and Borneo), in the central plain of Thailand and in the Pahang river of Malaysia. Maximum length recorded of the species is 50 cm. It is a colourful fish and the young ones (having a black lateral band) are used as an aquarium fish. In Thailand its reputation is not very high as a food fish but Malaysia it is a delicacy and the fish is sold at a very high price.

Jelawat is omnivorous in food habits and feeds on weeds, insects, molluses and small fishes.

At Batu Berendam Station, Jelawat culture was initiated in 1974, when 100 fingerlings (3"-4") were brought from Pahang R. The fishes have grown to only 1-2 kg size in 3-1/2 years. A second batch of 963 (1"-2") was exported from Indonesia through Singapore in 1976 and have grown to about 600 g. These fishes are now regularly fed with trash fish, chicken feed mixed with some weeds and also chicken insecticides.

Jelawat cultured in a private fish culturist's (Mr. Khoo Pak Wan) farm, visited by the consultant, have grown to the maximum size of 1.5 kg in one year. The fishes were fed with 80 per cent trash fish mexed with 20 per cent chicken feed. In Singapore, Jelawat fed with trash fish reported to grow to 900 g to 1 kg/yr.

For developing the cultural technique of Jelawat the fingerlings should be stocked at different stocking rated is fertilized ponds and fed with various types of supplementary feeds.

For brood stock development 40-50 larger specimens of 3-1/2 yr-old jelawat may be selected and stocked in Pond F12 (0.5 acre) provided with continuous flow of water and fed with 2 per cent of the body weight of feeds rich in animal protein, soyabean meal and some green leaves. The consultant examined the gut contents of a few jelawat which mainly consisted of weeds (Napier grass) and bivalves.

During the visit of the consultant, Jelawat (large size) ponds were netted and the fishes were examined. The weights varied from about 1 to 2 kg. Majority of the fishes appeared to be immature.

Twelve fishes selected for detailed study were anesthetized and detailed measurements taken and other studies made. Four specimens were sacrificed for internal study and the rest were released back into the pond. Of the four, one was a slightly oozing male and one female with slightly regressed gonad. The ovary had eggs of dark grey colour having an average size of 1.2 mm. The other two were immature females.

Sexual Dimorphism

Attempts were made to distinguish sex from the external characters of jelawat. There were no distinct roughness in the pectoral fin of the male or any tubercles present on the snout. Cloacal region also did not show much difference. However, the paired fins were observed to be comparatively longer in males than the females. In the mature males the first ray of the dorsal fin is distinctly more elongated than the female. Similarly, the anal fin in males appears to be comparatively longer than that of the female and reaches the base of the cuadal fin. This finding is only tentative and lias to lie confirmed by examining a good number of fully ripe male and female specimens. Both males and females had coloured fins in males appeared to have deeper orange tinge than those of females.

Induced Breeding

It was suggested that these maturing fishes should be examined occasionally and if found fully ripe, hypophysation may be tried in the same way as in case of *P. gonionotus* or silver carp/bighead. It was also suggested that someone should immediately be sent to Pahang river area where jelawats are available. In case information is received on the availability of mature specimens, a spawning team may go there, fix a breeding hapa in the river and give injections to the spawners. If this is successfully done then a stock of jelawat may be raised and further study on rearing of the larvae, fry and fingerlings can be initiated immediately.

Probarbus jullieni (Temoleh)

The species is available in the Perak and Pahang rivers of Malay Peninsula and also in the Mekong river in Thailand, Laos and Cambodia. It is the largest carp of Malaysia and is of high eating quality. The population in the local rivers had been greatly depleted. Probarbus jullieni was under investigation in a project supported by FAO/UNDP and some useful informations were obtained. The fish has a protactile mouth which suggests its feeding habits from the bottom mud. The food consists of aquatic worms, chironomid and insect larvae, bivalves, snails and plant materials. The fish spawns in rivers. Egg 2 mm in size is light creamy brown in colour. Induced breeding by hypophysation was previously tried but did not meet with complete success.

The consultant feels that considering the importance of the species and recognizing the fact that the population of the fish in natural habitat has been badly depleted, temoleh may be included for the study either in addition to or in place of *Tor tambroides/Puntius bulu*.

Following the induced breeding technique developed for Chinese carps this species could be attempted to spawn by hypophysation by obtaining ripe spawners from the Perak river above Chendroh Lake. The fry and fingerlings could be collected from the natural habitat and reared in ponds.

Grass carp (*Ctenopharyngodon idella*)

Project staff has more or less developed the technique of making grass carp breed by hormone injections. During the visit of the consultant all the three Chinese carp species were successfully bred by hormone injections. It would be easier to breed the 1976 induced bred stock of grass carps when they attain full maturity. More breeding experiments may be carried out for the standardization of the technique of hypophysation keeping proper controls. Data on the physico-chemical parameters, rate of fertilization and hatching rate etc. may be regularly recorded.

The institute has been using tannin for hardening the egg shell thus obviating the loss due to premature hatching because of the nature of water (softwater) used for the experiments. However, it has been observed that the hatching time is unusually prolonged to 24-30 hrs. in place of 15-18 hrs. This increases the cost in circulating the water in the hatching jars and also in loss of many hatchlings. Some experiments may be carried out for determining the optimum dose of tannic acid, the duration of the treatment, as well as the optimum time (hours after fertilization) when the treatment of eggs should he initiated.

The spawning team should also further standardize to determine the optimum number of eggs to be introduced in the hatching jar, the percentage of hatching and rearing of the post-larvae in the aquaria before release in the nursery ponds.

Pilot Hatchery for Production of Seed

A pilot fish hatchery has been designed for carrying out experiments on large scale production of the seed of the test species such as high bighhead and silver carp. Since the Engineers were working on the details, the consultant could not obtain the hatchery design for study. However, the project leader furnished the following information regarding number and size of spawning tanks, hatchery jars and water requirement for the hatchery.

When the entire hatchery is operational, then total water requirement will be 52,445.35 cft. This quantity is 10 per cent of the total volume of A1 pond (4 1/2 acre) when full.

(a)	Size of pilot fish hatchery 120' x 40'	
(b)	Spawning tanks with showers [circular with collection chamber as used in People's China]	
	(4 nos.) 9' diameter x 40" deep Collection chamber ----- 4'7" x 6' x 40"	21,641.76 cft
(c)	Smaller spawning tank (10 nos.) 6' x 3' x 2'	13,608.00 cft
(d)	Spawnery (3 nos.) 10' x 5' x 2'	2,289.60 cft
(e)	Hatching jars (14 nos.)	8,898.50 cft
		47,097.36 cft

The water for the hatchery would be pumped in from A1 pond by the side of which the hatchery is planned to be constructed. According to the plan the water from the hatchery would be returned back to the same pond. The consultant suggested that instead of returning the used water direct to the same pond, it could either be diverted to the adjacent pond or the used water flown through open channel and released to the distant end of the pond.

With the successful operation of the hatchery, live feed culture system has to be further intensified. More tanks of large size (1/2 to 2 ton) may be required for culturing live feed for post-larvae and fry.

Sufficient supply of quality water for the hatchery and the experimental ponds may be a problem in near future in the Freshwater Station, Batu Berendam. During the period of visit of the consultant the pump station at the intake site was out of action. Since water level in the river had receded further, the intake pipe had to be further extended. The socalled Malacca river appeared to be in the process of being converted into swamp. The whole surface is covered with marshy weeds and there is practically no movement of water in the river. Besides, the water also receives discharge from some factories. Unless an alternative source of water is found out this water might create problems in the efficient working of hatchery. Alternatively, provision for recycling the water back into the hatchery system by passing through filters should be made.

Nursery and Rearing Pond Management

Management of nurseries and rearing ponds is very important for obtaining better growth and higher survival of fry and fingerlings. FFRS has the advantage that the farm ponds can be drained out completely and refilled.

Weed Control

Weed eradication in small nurseries is not a problem. This can be done manually. But in larger rearing ponds the submerged weeds and filamentous algae can be biologically controlled by grass carp. The floating weeds, excepting *Eichhornia* and *Pistia*, can be efficiently controlled by grass carp. Excessive bloom of blue-green algae is rather a problem in intensively cultured ponds. Microcystis bloom can be suitably controlled by application of Simazine but fishes get killed when depletion of O_2 occurs due to rotting of the dead algae. An impending bloom of Microcystis or Anabaena can be eliminated by diluting the pond water thereby changing the ecological condition of the pond.

Control of Aquatic Insects

Air breathing aquatic insects (Notonectids etc.) can be controlled by treating the pond surface with soap-oil emulsion (vegetable oil 50 kg/ha mixed with cheap soap 17 kg/ha). However, the expenditure can be reduced if low speed diesel oil is used with a suitable detergent. The larvae of diving beetles are highly predatory. Their occurrence can be prevented by keeping the marginal area of the pond cleared of weeds because the beetles lay their eggs on the stems of the marginal weeds. Small insects like notonctids may not be a big problem if the post-larvae are reared up to 13-14 mm. size.

Stocking of Nurseries

Stocking rate of post-larvae is very important. When nursery ponds are stocked with high rate of stocking such as 3.75 m/ha, care should be taken to produce sufficient zooplankters in the pond by proper fertilization. In addition, the fry should be supplied with supplementary feeds such as rice bran, mixed with finely powdered peanut meal or soyabean meal or any other suitable substitute. Addition of a very small quantity of growth-promoting substances, such as $CoCl_2$ (micronutrient) with the supplementary feed may help in obtaining better growth and higher survival of fry.

One of the most important measures in nursery management is thinning out of fry in appropriate time. The fry should not be kept in crowded condition for a long time. It is advisable to rear post-larvae for 2 to 3 weeks during which the fry is expected to grow to 25-30 mm and then thinned out and stocked in rearing ponds for growing those to fingerlings. Rearing pond management is more or less similar to nursery management.

During the visit of the consultant demonstrations in the ponds could not be given as intended since the ponds if drained out could not be refilled or replenished with water because the intake pump was out of action.

Study on Nutritional Requirements of Carps

This important aspect of study appears to get a further set back in Phase II of the project. As of now, the Project Adviser (Nutrition) has left and the counterpart officer is away on training and he may not be back till the middle of 1979 (as reported). Mr. Hanif is working on this project and is trying his best to formulate some diet from the locally available ingredients. Although he is an enthusiastic researcher, he has no previous experience and training in this field of study. As such, hardly any substantial progress on this important aspect of the research programme could be expected till Mr. Pathmasoothy returns to Malacca after completing his training.

Preliminary Investigations on Some Selected Natural Impoundments of Stocking those with Fry and Fingerlings

One of the objectives in Phase II of the project is to make preliminary investigations on selected natural bodies of water for stocking fry and fingerlings of selected species. Studies on the physical, chemical and biological parameters of the natural water bodies or impoundments may be made initially to determine their suitability for stocking. Study of primary productivity may be initiated by 'light' and 'dark' bottle technique or by C_{14} technique for the determination of carbon assimilation by the planktonic algae which gives a measure of organic production. Based on the productivity of the waters the suitability of species to be stocked can be determined. Trial stocking of selected species in floating cages and/or pens may be done next. While water rich in zooplankters could be stocked primarily with bigheads, impoundments predominant in algae could be stocked mainly with silver carps. A weed-infested (submerged) waiter could be ideal for grass carp. However, sufficient precaution, should be taken before introduction of any species, and study made on the fish population present in the water bodies both qualitatively and quantitatively and efficiency value of each species determined.

Some preliminary observations have already been initiated by the project staff in a nearby reservoir Durian Tunggal. The consultant visited the reservoir which was filled in Feb. 1977 and has a water area of 480 ha. The average depth of the reservoir is about 15 m with the maximum depth of 27 m. The water above the dam had a light bottle-green colour indicating the water to be fairly productive. This was confirmed by the physico-chemical parameters studied at the time of the consultant's visit. pH of water recorded was 7.3 and dissolved oxygen was optimum. The plankton collected by sieving 45 litres of water was examined by the consultant and was found to be fairly rich. It consisted of Cyclops, Brachionus, Tetramastix, Ostracods, Phacus, Scenedesmus, Cosmarium, Naviculids, and very minute unicellular green algae. Cage culture in deeper waters and pen culture in shallower fringe areas may be tried. Since both zoo and phytoplankters are present both bighead and silver carp can be tried at different proportions and densities. In this connection it may be mentioned that cage culture with bighead and silver carp have given very high production of 25 kg/cu m/10 months in reservoirs in Singapore when stocked at the rate of 10 fingerlings (15cm.size) of bighead/silver carp per cubic meter.

The consultant also visited a 0.75 acre pond belonging to Paya Lang (Dunlop) Rubber Estate and a 40 acre reservoir in the Batu Enam Estate (Harrison and Crossfields). The water in both these bodies of water was acidic (6.6-6.7) as was obvious from the rusty colour of water.

Cage culture of *Tilapia mossambica* in the reservoir was suggested. Tilapia in floating cages usually cannot reproduce since nest making is not possible.

Pituitary Bank

The common carp *Cyprinus carpio* is an ideal donor fish because of its comparatively larger size of pituitary gland and is easier to collect and also for the availability of mature common carp for almost throughout the year. Experiments have already been initiated at FFRS in the production of common carp as donor fish in four 0.5 acre 'E' ponds. When sufficient number of donor fish is raised in the farm the cost of pituitary glands would be negligible since the fish could be sold after collection of the glands. Effectiveness of pituitary glands under various methods of preservation will be studied. Pituitary extracts made in glycerine and water (1:2) and ampouled can remain effective for a longer period. Preliminary work carried out at FFRS gave satisfactory results with pituitary glands when donor recipient was in the ratio 2:1. Further work in this line will be carried out such as, correlation of body weight of fresh pituitary gland and correlation of weight of alcohol preserved gland to the weight of fresh pituitary gland. A bioassay technique will be developed with various pituitary materials.

Simultaneously, the grant-in-aid project with Universiti Sains Malaysia would be engaged in trying alternate donor species such as, Rastralliger and Decapterus because of non-availability of good tuna donors. The bioassay technique would be attempted on common carp as the test animal. Tawes (*P. gonionotus*) may also be a good test animal for such studies. A simple bioassay technique is being developed by Dr. Khoo Khay Huat at the Universiti Sains Malaysia.

Transport of Fry, Fingerlings and Spawners

Transport of fry and fingerlings under oxygen packing has been successfully carried out in many countries. If proper care is taken in handling the fish and they a reconditioned for several hours with intermittent splashing of water (for emptying the stomach), long distance transport can be made without any mortality. However, experiments have to be conducted to determine the optimum number to be transported depending on the size of the fish and duration of transport. During the consultant's visit a consignment of advanced grass carp fry were conditioned and transported in plastic bag under oxygen packing. For the transport of large fishes or spawners, it may be necessary to use anesthetics in the water. For transport involving 4 to hours duration, anesthetic may not be necessary if the fishes are transported in canvas tank under oxygenation. At Pandan and

Further refinement and standardization of the technique of spawning grass carp may be attempted in Phase II project using HCG in combination with carp pituitary gland and by regulating the frequency of injection and interval between subsequent injections. For synchronizing optimum time for stripping, the progress of hydration in the injected female may be closely watched. The synthetic analogue of the nonapept de LH-RH found highly effective in induction of spawning in Chinese carps in People's China may be tried in grass carp.

Hatchery techniques may be further refined with a view to obtaining higher rate of fertilization, hatching and higher survival of post-larvae. The optimum dose of tannic acid, duration of treatment and the stage of development of eggs at the time of treatment may be determined by repeated trials so as to cut down the long extended period of incubation and obtain better hatching.

Sufficient care may be taken in the preparation of nurseries so as to ascertain that no extraneous predatory or weed fishes get into these. Further experiments on the fertilization of the ponds preferably with a mixture of suitable organic and inorganic fertilizers may be tried. Once fry are introduced, use of organic manures for replenishing the depleted live feed is risky. In such case, intermittent low dose of inorganic fertilizer such as urea is advisable.

Soap-oil emulsion is useful for eradicating backswimmers and other smaller air breathing insects. In addition, it is advisable to net the pond several times with a close-meshed net for the eradication of miscellaneous predatory insects. When the stocking rate of fry in fertilized nurseries is very high (>3 m/ha) supplementary feeding with finely powdered suitable feeds if, essential for obtaining high survival. Addition of growth-promoting substances like $CoCl_2$ with feed helps better growth and higher survival. Thinning out of the crowded population is absolutely necessary in appropriate time otherwise heavy mortality may occur. It is suggested that the post-larvae may be grown in a nursery for 2 to 3 weeks (25-30 mm size) after which the fry should be transferred to rearing ponds at a much lower rate of stocking.

A pilot fish hatchery is under construction at FFRS at a cost of M\$ 100,000. The circular spawning tanks with chamber have been designed on the model used in People's China. The source of water is one of the 4 1/2 acre ponds. Success of the hatchery will greatly depend on the quality of the water. In future, it may be necessary to provide facilities, for recycling the water after passing through efficient filters so as to maintain the supply of quality water to the hatchery.

The project staff has developed a system of live feed culture. With the establishment of the fish hatchery the facilities for live feed culture have to be expanded considerably. It may be necessary to have pure culture of Chlorella and other suitable algae for supplying feed for the mass culture of rotifers, cladocerans and copepods.

Culture of common carp as a donor fish has recently been initiated. Since common carp is an ideal donor fish, large-scale culture of the species fertilized ponds with supplementary feeding may have to be developed, keeping in mind the danger of destruction of the pond embankment by the species. Since the pond embankments are protected by packing of car tyres, the common carp might not be able to destroy it. A bioassay technique with the carp gonadotropin may be developed in due course.

Study of the examination of various impoundments are natural bodies of water for stocking of selected fish is being initiated at the beginning of Phase II of the project. Study of the primary productivity of such waters may be made by using either 'dark' and 'light' bottles or C_{14} technique so as to get an idea of the organic production of the water. Floating cage culture in deeper water and pen culture in the fringes with suitable exotic and local fishes, depending on the type of food available, will give an idea of the suitability of the species to be stocked. However, before initiating any stocking programme the existing stock of fish may be ascertained both quantitatively and qualitatively.

Index